Praise for *In Quest of a Shared Planet*

"If there's one country on earth that has the most at stake in slowing climate change, it might be Bangladesh. So it makes great sense to hear the story of the global climate negotiations from this perspective—it will be of interest to anyone who has followed these talks, or who wants to understand how the world looks different depending on where on it you were born."

—Bill McKibben, author *The End of Nature*

"An outstanding book, by an excellent scholar writing in a popular voice. The book is a crucial resource for those seeking to understand the COP process, particularly those who are planning to attend as delegates."

—Jessica O'Reilly, Indiana University

"This is a fascinating and unique book. So much has been written about the success and failures of the international climate negotiations by political scientists and by Northern analysts. Khan comes at the question entirely differently. As an anthropologist, she follows Bangladeshi diplomats, analysts, academics and activists to understand what draws and keeps people within the tortuous negotiating process. Her answer will surprise you."

—Mike Hulme, University of Cambridge

"*In Quest of a Shared Planet* is a highly original account of the climate negotiation process, written in a refreshingly personal style. Khan's book works through the difficult issues at the center of why humanity has not successfully dealt with climate change through UN-led negotiations. Khan hammers home the importance for developing countries of issues like payments for damages they'll experience from climate change they didn't cause."

—J. Timmons Roberts, Brown University

In Quest of a Shared Planet

NEGOTIATING CLIMATE FROM THE GLOBAL SOUTH

Naveeda Khan

FORDHAM UNIVERSITY PRESS   NEW YORK   2023

This book is freely available in an open access edition thanks to TOME (Toward an Open Monograph Ecosystem)—a collaboration of the Association of American Universities, the Association of University Presses, and the Association of Research Libraries—and the generous support of Johns Hopkins University. Learn more at the TOME website, available at: openmonographs.org.

Fordham University Press has no responsibility for the persistence or accuracy of URLs for external or third-party Internet websites referred to in this publication and does not guarantee that any content on such websites is, or will remain, accurate or appropriate.

Fordham University Press also publishes its books in a variety of electronic formats. Some content that appears in print may not be available in electronic books.

Visit us online at www.fordhampress.com.

Library of Congress Cataloging-in-Publication Data available online at https://catalog.loc.gov.

Printed in the United States of America

25 24 23   5 4 3 2 1

First edition

To Sophie, Suli, William, George, Faizah, Ameera, Zoha, Faraz, Ayan
This is to help you in whatever small way to fight for your future.

Contents

Acronyms and Abbreviations

ADP	Ad Hoc Working Group on the Durban Platform for Enhanced Action
AGN	African Group of Negotiators
AOSIS	Alliance of Small Island States
APA	Ad Hoc Working Group on the Paris Agreement
BASIC	Brazil, South Africa, India, and China
CAN	Climate Action Network
CBDR	Common but Differentiated Responsibilities
CJN	Climate Justice Network
CMA	Conference of Parties serving as the meeting of the Parties to the Paris Agreement
CMP	Conference of Parties serving as the meeting of the Parties to the Kyoto Protocol
COP	Conference of Parties
CSO	Civil Society Organizations
CVF	Climate Vulnerable Forum
DCJ	Demand Climate Justice
ECBI	European Capacity Building Initiative

ENB	Earth Negotiations Bulletin
ENGO	Environmental Nongovernmental Organizations
FOE	Friends of the Earth
FOEI	Friends of the Earth International
G-77	Group of 77
ICCCAD	International Center for Climate Change and Development
IIED	International Institute for Environment and Development
IISD	International Institute for Sustainable Development
IMF	International Monetary Fund
IPCC	Intergovernmental Panel on Climate Change
LDC	Least Developed Countries
LMDC	Like-Minded Developing Countries
MEA	Multilateral Environmental Agreement
NDC	Nationally Determined Contributions
NELD	Noneconomic Loss and Damage
NGO	Nongovernmental Organizations
OECD	Organization for Economic Co-operation and Development
OHCHR	Office of the United Nations High Commissioner on Human Rights
OPEC	Organization of the Petroleum Exporting Countries
PAWP	Paris Agreement Work Programme
PKSF	Palli Karma-Sahayak Foundation
QUNO	Quaker United Nations Office
RINGO	Research and Independent Nongovermental Organizations
SBI	Subsidiary Body for Implementation
SBSTA	Subsidiary Body for Scientific and Technological Advice
SIDS	Small Island Developing States

TUNGO   Trade Union Nongovernmental Organizations

TWN   Third World Network

UNFCCC   United Nations Framework Convention on Climate Change

UN   United Nations

UNDP   United Nations Development Program

UNHCR   United Nations High Commissioner on Refugees

WIM   Warsaw International Mechanism for Loss and Damage

WMO   World Meteorological Organization

Bodies under the UNFCCC

Supreme

COP, CMP, CMA. Make global policy relating to climate change, under the UNFCCC, Kyoto Protocol [CMP], and Paris Agreement [CMA], respectively. They either approve conclusions of discussions in sessions or adopt decisions, which are binding.

Subsidiary

SBSTA. Permanent body that provides the scientific and technological advice needed to support the supreme bodies; provides linkages between other UN entities, such as the IPCC and the UNFCCC process.

SBI. Advises on means of implementation, monitoring and review processes, and provides education and training.

Constituted bodies

Such as:

Adaptation Committee

Consultative Group of Experts

Executive Committee for the Warsaw International Mechanism for Loss and Damage

Least Developed Countries Expert Group

Standing Committee on Finance

Created at different points in the process by various COP decisions; time bound, although open for negotiated renewal.

Funds and financial entities
Adaptation Fund
Global Environment Facility
Green Climate Fund
Least Developed Countries Fund
Special Climate Change Fund
Created at different points in the process by various COP decisions; funding
    sources not secured and always up for discussion; the means by which fi-
    nance is made available by developed countries for developing countries.

Bodies that are mentioned in the book and have concluded their work
Ad Hoc Working Group on the Durban Platform for Enhanced Action (ADP)
Ad Hoc Working Group on the Paris Agreement (APA)

IN QUEST OF A SHARED PLANET

Introduction

The Climate Regime

"Is it going to solve climate change?" This is the first question I get when I say I've been studying the United Nations–sponsored global climate negotiations since the crafting of the Paris Agreement in 2015. Given that anthropogenic climate change is already here and "baked into the earth system," the problem isn't going away any time soon (Zhou et. al. 2021). At best, our actions today can ensure its alleviation, perhaps a reversal of some of its worst impacts, fifty to a hundred years from now. When we hear that we have only a few decades before the window of opportunity closes on meaningful change, that means that if we don't act decisively now, we can expect the average global temperature rise to soon reach 1.5 degrees Celsius above preindustrial levels. Report after report has already told us what we stand to lose of our world with every degree the Earth warms (Lynas 2008).

One glance at the timeline of the global climate negotiations and it is clear that the UN process hasn't delivered decisive action and won't anytime soon. The process started with a bang in 1992 with the Earth Summit in Rio and a robust convention, the United Nations Framework Convention on Climate Change (UNFCCC), but only now, almost thirty years on, do we have a collectively agreed-upon approach to the problem. The Paris Agreement, which effectively replaced the 2007 Kyoto Protocol, was signed and ratified in 2015. Although a majority of its rules for implementation were adopted in 2018, important elements of the agreement continue to be negotiated.

The first global review of what the Parties to the Agreement—the convention's term for its signatory nations—have achieved since they last provided reports in 2015 will happen in 2023. This review will indicate how much the countries of the world have achieved toward keeping temperature rise to

2 degrees Celsius compared to preindustrial times, a goal written into the Paris Agreement, although 1.5 degrees remains an aspiration. It will be followed by periodic country reports submitted every five years and stocktaking also every five, detailing our slow, incremental battle against a fast-spreading and heterogeneous problem.

In response to the question of whether the negotiations will solve climate change and the attendant anxiety-producing question of whether it will save humanity and biodiversity, I venture to say that we won't see a solution through this process, at least not in the form or at the pace at which it is occurring. Many people involved in the negotiation process say as much. They point to the increase in extreme climate events and infrastructure failures, compounded by ineffective governments. They say the Paris Agreement is too little, too late. Many wonder if it is time to abandon ship and retreat to national, subnational, and local efforts, holding out hope for a bottom-up mode of self-organizing rather than a top-down approach.

Regardless of their disappointment and growing skepticism, many return year in and year out to the annual UN sessions to pore over documents, attend endless meetings, see countless people, advocate for policies, and communicate what they are doing to wider constituencies. Of course, for many, attending these sessions is their profession, but they do not go about it as if it was merely that. They treat it as a vocation, even a calling.

Persuaded by their tirelessness, I decided to throw myself into studying the process. I didn't do it to eke out hope for a way out of our current conundrum. My goal was to understand what keeps people in the process. But something happened along the way. I developed a stake in the people and, as a result, in the process. I came to feel that we might just accomplish something through it, provided we keep watch over it, prod it to be more inclusive, more generous, more ambitious, while being more honest about its limits. It was that or let it devolve into the grotesque masquerade it sometimes becomes. This is a story of my slow conversion to the process and my dawning sense of what it might accomplish, and it is an invitation to others to join in making the process be its best.

Let me just say off the bat: The process is mammoth, with different legal agreements, bodies, meetings, sessions, decisions, workstreams—and acronyms for them, lots and lots of acronyms. It was not possible for me to cover even a representative sample of the individuals involved. I decided early on to privilege some over others. Given that I entered the process through my field research in Bangladesh, I focused on Bangladeshis of all stripes. Given that Bangladesh is securely in the world of have-nots and affiliated with developing countries, I decided to privilege countries of the Global South, so-called for being at the periphery of the industrialized North, who seek to constitute a solidarity group in international arenas on issues from trade to the environment.

Traversing the many domains within the process—activism, civil society, country delegations, the political coalitions of the Global South, technical expertise, UN officialdom, media, and academia—I asked people over and over why they were there and what they hoped to gain. I was treated to a wide range of replies. To my surprise, the responses did not engender cynicism toward the process or make me bemoan the contemporary global political system, with its sharp divide between the haves and the have-nots, a divide that was invariably part of and structured the process (see Corporate Accountability 2019). Instead, it gave me an appreciation for the effort people put into continuing to be in conversation, despite, or maybe because, of the unequal world.

At our most charitable, we can say the UN-sponsored global climate negotiations showed themselves to be about keeping countries, which were often antagonistic, together in conversation, to allow the realization to dawn that they share a planet, despite deep inequalities, different existential realities, and serious obstructions to mutual understanding. The climate negotiations gave some a political voice or a moral authority that they lacked in other international spheres. While references to violent pasts and histories of colonialism, slavery, and extractivism that had spurred industrialization in specific regions of the world at the expense of others were sparse within the negotiations, it was still a space in which the past of industrialization could, within limits, be brought up to parse historical responsibility for the way the planet is at present. The very fact of anthropogenic climate change made it impossible to sweep industrialization under the carpet.

For others, this congregation was an opportunity to access new networks, funds, and projects. An even more uncharitable reading of the process, espoused by many, was that it was the way for the Global North to distribute the burden of fighting climate change over the entire globe, including to those who had not contributed to creating the problem.

It was Doreen Stabinsky who brought home to me the necessity of the climate change negotiations and the fact that the process may yet deliver more than it promised. Doreen was a professor of global environmental politics at the College of the Atlantic, in Bar Harbor, Maine. She recounted that she became involved in the UN process some twenty years ago, as a member of civil society, attending out of a sense of responsibility to monitor and put pressure on issues related to agriculture and food security. Over time, her politics became more far-reaching. She became less interested in how to ensure food security and more interested in food sovereignty, that is, the right to food and the right to define the terms of a sustainable food system. She became more involved in helping countries from the Global South navigate the Conference of Parties, or COP (which confusingly refers both to the supreme decision-making body of the convention made up of signatories to it and to the annual meeting of the body), providing them with the knowledge and expertise she had gathered from her years of attending COPs. She brought students from her college to the negotiations, and they became useful aides to countries from the Global South, which lacked the manpower to attend concurrent meetings across all the issues under negotiation.

Over time, Doreen became more and more interested in the issue of loss and damage. When neither the issues of mitigation nor adaptation seemed to be making much progress within the process, the issue of loss and damage emerged. I explain what these climate actions mean later in the book, but suffice it to say that loss and damage was a contentious issue on which there was no consensus but which surprisingly found a home within the Paris Agreement. At present, the dangers posed by climate change were assessed in terms of risks. But as these dangers began to actually manifest in outright destruction, questions of who was responsible, for what, and to whom would arise, raising further questions of liability and compensation. Consequently, the inclusion of loss and damage in the agreement made countries of the Global North nervous, lest the formulation one day attain the political and legal clout to swerve the process toward litigation to force climate actions, rather than negotiated policies and practices. Doreen was among those who supported and pursued the inclusion of loss and damage within the agreement.

During one of our conversations, I asked her why she devoted so much time to this one issue, since the idea of loss and damage always seemed to be at the verge of extinction, given the intensity of opposition to it from developed countries. Her response was illuminating. Doreen said that she felt it necessary for people to put in the labor of figuring out terms, hashing out definitions, putting in place procedures, collating best practices, and undertaking legal actions, even with no likelihood of swift judgment, because climate change would in any case bring about the erasure of our inherited concepts in favor of new ones (Stabinsky and Hoffmaister 2015). In this (near) future, loss and damage would be an ordinary occurrence, governments would be made unstable by it, and countries would be in a constant state of crisis. Putting in the effort now could provide the template for a shared global understanding and protocols for action when everything else failed.

Whether or not we subscribe to Doreen's view of the end of the known world and the rise of a radically different one, she provided me with a sense of how the process had the capacity to hold together both the utterly banal and procedural and the urgency of the sense of the end of days. She led me to think that one could both diagnose the problem with global politics and produce a new politics or at least new political norms. The objective of the Paris Agreement was to evolve, slowly but steadfastly, new ways of interrelating and being accountable to one another.

The UN-sponsored climate negotiation process must be distinguished from the United Nations. Although it was highly dependent on UN codes of conduct, modes of procedure, and even funding, it was mostly independent of the UN at this point. As my interlocutors said, this process was now almost entirely Party-led, that is, a country-led political process. Although there were many possible starts to the UNFCCC process, such as the 1972 United Nations Conference on the Human Environment in Stockholm, which led to the formation of the United Nations Environment Program, the legal scholar Daniel Bodansky (2001) makes the important observation that those early efforts at addressing the environment mainly focused on pollution, specifically the reversible effects of pollution such as oil spills, hazardous wastes, and air quality. The more significant environmental efforts leading to the UNFCCC process focused on long-term, potentially irreversible effects, which came into view only after the integration of global data.

As Paul Edwards has shown in *A Vast Machine* (2010), it took considerable coordination of real-time experiments, data collection, models, and simulations to bring together disparate understandings of the weather, regional variations, and the different spheres (biosphere, atmosphere, lithosphere, hydrosphere), along different timelines, to allow a picture of the climate as a global system to come into focus. As both Bodansky and Edwards have observed, the science of climate change fed directly into advocating for social and economic policies to abate climate change and the political process needed to bring about such policies.

The frontrunner of the climate process was the international and national mobilization to counter the depletion of the ozone layer, which led to the passing of the 1987 Montreal Protocol on Substances That Deplete the Ozone Layer to control for gases deleterious to the outer atmosphere.

The World Meteorological Organization and United Nations Environment Program–sponsored Intergovernmental Panel on Climate Change (IPCC) emerged in 1988 not only to confirm the scientific consensus on climate change but to create political consensus across nation-states on the reality and extent of climate change. It released its first assessment report in 1991. Meanwhile, the UN-commissioned 1987 Brundlandt Report, *Our Common Future*, brought the environment together with other concerns to argue for the necessity of sustainable development, that is, economic growth that kept the environment, limited resources, and future generations in its view.

When the sovereign nations of the world congregated in Rio de Janeiro for the United Nations Conference on Environment and Development (also called the Rio Summit or the Earth Summit) in 1992, they were well primed on the issue of climate change, as distinct from environmental pollution. They negotiated and signed three treaties, the United Nations Framework Convention on Climate Change (UNFCCC), the Convention on Biological Diversity, and the United Nations Convention to Combat Desertification.

The UNFCCC, also referred to as the Convention or FCCC, was ratified in 1994. In international law, a convention is a broad-based international commitment to an issue. It presumes later sessions, negotiations, agreements, and national legislation to realize this commitment in actionable terms. Its very existence calls into being a process. The UNFCCC gave rise to such a process, starting with the formation of a secretariat also, confusingly, called the UNFCCC or, sometimes simply the Secretariat, to provide organizational and technical support to this process. The Secretariat was based in Bonn, Germany, and helped countries host the annual COP. The COP had been taking place since 1995 and had yielded many decisions, bodies, funds, and programs in the twenty-four some years since its establishment. Most notable among them were the two multilateral environmental agreements, the 1997 Kyoto Protocol and the 2015 Paris Agreement.

The hope was that the UNFCCC process would yield a global plan of action for combating climate change, like the 1987 Montreal Protocol. The Kyoto Protocol was the first attempt at such a plan, but it failed for reasons that will be discussed in this book. The Paris Agreement was now widely viewed as the plan of action to which almost all the countries of the world agreed. However, the years it had taken for these negotiations to evolve into an agreement showed that the climate was not quite the same as the outer atmosphere and that there wasn't a single approach to the problem of climate change, as there had been to holes in the ozone layer.

The process of addressing the changing climate had not been a linear unfolding from identifying the problem to agreeing on its possible solution but rather an aggregate of processes, each with its own promises and time horizons. It was composed of the convention, a clutch of environmental agreements, annual Conferences of Parties and intersessions, attendant subsidiary bodies, and funds, backstopped by the UNFCCC Secretariat and IPCC. While it might not seem like much, as an interlocutor said, "Unlike the UN, the UNFCCC is a process." In other words, it was still evolving to meet the challenge set it, and once it had met its mandate, it would become redundant.

I suggest that while the climate had indeed produced a sprawling process, which fell under the umbrella of the UNFCCC, this process was far from ecological, if by that we mean that the parts were intrinsically interrelated and interdependent, as was sometimes claimed of the UN system or the international community of nation-states. Regardless, the process, such as it was, afforded a playing field for countries of the Global South in which the terms of participation were somewhat more favorable than other fields, such as the global economy. For instance, here developing countries were not the debtors that they tended to be in other arenas but to whom something was owed.[1] In the mainstream of UN discourse, there tended to be an acknowledgment that developed countries had a greater responsibility for mitigating climate change because they had more money, capacity, and technology to do so. Developing countries were empowered to request finances, enhancements of their capacities, and technology transfers to better fight climate change without it immediately seeming like an effort at resource redistribution, which historically had been feared and blocked by developed countries (Getachew 2019).

In the climate regime, you see not only the points of tension between the Global North and the Global South but also differentiation within the Global South, leading to what Gregory Bateson (1958) has called "schismogenesis," the routes by which divisions lead to disorder and incoherence until such time as existing links are broken and new ways of communicating and exchanging information arise. Schismogenesis may have haunted the countries of the Global North as well, but I made a decision not to study the Global North to the same degree. My previous research in Bangladesh shaped my commitments toward those less empowered and less heard within negotiations, even if it meant ultimately complicating the Global South as a coherent formation (see Prashad 2013).

Through a focus on Bangladesh, I explore how small, poor countries deployed not just soft power or moral authority, both of which have been written about in the field of international relations (Marquardt 2017; Genovese 2020), but rather weak ontology, by which is meant a self-doubting wager of sorts that evolves its own modes of self-affirmation (see White 2000), such as we see in Bangladesh's fight to retain the notions of loss and damage within the negotiations. I show how hard it was to maintain this kind of existence within the process, how full of compromises it was, fringed as one's existence was by developed countries and large developing countries, with history and the past written out of the process and a certain presentism placing a stranglehold on imaginations of the future (see Paprocki 2021, 2022). The figures of myriad civil society participants, activists, and foreign and in-country experts fade in and out of view as they attempted to influence national policies and politics, with various degrees of access and success. Over the course of this process, we see issues close to the heart of the Global South—equity in its various forms—transmute, becoming shadows of their previous selves as mitigation took on more and more specificity through metrics and standards.

The Paris Agreement, though not as accommodating of the circumstances of developing countries as the original 1992 UN climate convention or the Kyoto Protocol, which had initiated this process, nonetheless still held out hope for change through the absorption of new norms of mutual accountability.[2] What finally emerges out of my exploration of this cross-section of the process was the importance of the youth movement for climate, which I claim was produced from within this system and spurred by its failures. The youth movement's critique of its elders was made poignant through their solidarity work with smaller, poorer countries. The process could not shake them off or treat them as pesky outsiders who did not understand the dynamics of multilateral diplomacy, because their critique was leveled from inside.

In studying the climate negotiations, others have written from the standpoints of international studies, law, and political science—fields for which countries exist as sovereign entities and operate like persons engaged in multilateral diplomacy within the global political order (Kaufman 1988; Boisard, Chossudovsky, and Lemoine 1998; Walker 2004). Studies such as Steven Bernstein's *The Compromise of Liberal Environmentalism* (2001) or Joanna Depledge's *The Organization of Global Negotiations* (2013) further home in on the process of producing multilateral environmental agreements, providing useful historical accounts and proposing theories that might predict outcomes. While I drew on this rich archive of frameworks, terminology, and analysis, I also wrote this book as an anthropologist. This means several different things. I reached for anthropological works on international organizations and law. Among the scholars who have shaped this field are Annelise Riles, Sally Engles Merry, and Ronald Niezen and Maria Sapignoli. These scholars allowed me a transverse view of the process through their insights into how different bodies of law and political norms awkwardly abut one another (Merry 2006, 2007, 2011). These works drew my attention to the unstated formal and aesthetic elements of international law and legal procedure in addition to their substantive content (Riles 1999a, 1999b, 2000), a method I extended to the study of the Paris Agreement. And they insisted on attending to the historical conditions, social practices, and aspirational and affective dimensions of any process or organization (Niezen and Sapignoli 2017), an orientation that runs through this book.

I also reached for concepts forged from research in different parts of the world that might seem foreign to official UN institutions and meetings. For instance, I employed the metaphors of sorcery and possession to explain how one comes to be seized by the process, compelled to speak only through the terms it gives (Favret-Saada 2015). I drew on the literary theorist Mikhail Bakhtin's notion of polyphony to show how Bangladesh Party delegates spoke in different voices, and I used ordinary language philosophy to ask whose voices they ventriloquized (Cavell 1994). Anthropological modes of attention to texts helped me read the Paris Agreement for the forms and norms of the genre (Barber 2007). I have already spoken of the concepts of schismogenesis and weak ontology in describing the behavior of nations and their bases of authority. I explored how, even if they represented specific countries, many treated the process more as a kind of vocation, in Weber's (2004) sense, as an end in itself rather than an instrument for diplomacy. I plumbed the aspect of reciprocity that anthropology has long studied to ask whether there was scope to go beyond debt and repayment and imagine the process within an economy of gifts (Mauss 2002; Hénaff 2019). The anthropology of climate change was a rich subfield that informed my work, ranging from knowledge production to the variegated experiences of the local impacts of climate change (see O'Reilly et al. 2020).

Being an anthropologist also meant that I spent a lot of time attending sessions. Initially, I did what is colloquially referred to as "deep hanging out," to get a feel for the process. Later, I carried out more systematic research, moving from participant observation and corridor conversations to formal interviews, archival research, and textual analysis, to cover the areas I had marked out as both interesting and relevant to the story of the process and its fluctuating promise for its participants. My writing is informed by anthropology's commitment to ethnography, systematic descriptions of a group of people but without claiming to provide a bird's eye or objective view of the whole. I used myself as a gauge for first impressions, feelings of being lost and, over time, growing understanding of the patterns around me before attempting to gauge how people sought to shape the future.

I deployed anthropology to make the process, which was slowly becoming familiar and commonsensical to me, simultaneously unfamiliar to my readers. Prediction generally relies upon building a model of what is. That is, it gives a lot of stability to the present to account for what will be. Anthropology, on the other hand, eschews models for description and diagnosis to make that which is seem less stable, inevitable, and predictive of the future, brimming instead with alternatives and allowing for the possibility of surprise and change. My hope in doing this was to allow the rich archives on the diversity of human existence and relations to open new possibilities for politics within the existing global political order instantiated by the UN climate negotiations.

In the pages ahead, the reader will find descriptions of my repeat attendance at the Conference of Parties over a five-year span (COP21 through COP25), my exploration of meeting spaces, and my growing understanding of the climate negotiation process. With a focus on the Bangladeshi delegation at the COPs, I draw out what it meant to be a small, poor, and dependent country within the negotiation process. I speak with negotiators within country delegations to explore their pathways to the negotiations and how they viewed negotiation as their vocation. I observe training sessions of negotiators of the Global South to draw out the understandings of what is or is not achievable within negotiated texts and the power of deal making and deferrals. I track the many-striped experts who provide guidance to delegations from the Global South to show how they attempted to influence the negotiation process. I profile individuals who had committed themselves to the climate negotiation process, moving between the Secretariat, Parties, activists, and the wider UN system to bring their principles, strategies, emotions, and visions into view. Culling insights from lawyers, I compare the discursive thrust and motivations for the Preamble and articles of the Paris Agreement (PA) with the Kyoto Protocol (KP), which the PA effectively replaced. I unfold the negotiation over one of the most contentious tools by which the Paris Agreement was to be implemented, that is, the Nationally Determined Contributions (NDCs). I explore how the newest pillar of climate action, loss and damage, emerged historically and how developed countries within the process attempted to control it, and I show how loss and damage ramified in many different and unexpected directions. I suggest that we understand the Global South's pursuit of loss and damage as not only a politics of forcing the issue of a conjoined future upon the Global North but as a gift to the youth of the Global North to secure that future.

With this book, I hope to give the reader a first-person perspective on this process to help them understand it, navigate it if they plan to attend themselves, and critique it—but with sympathy and, I hope, with an eye to alternatives. It also aims to introduce students of environmental issues to the only existing global process on climate change that exists at present. And it hopes to recall an older way of doing politics through tenets of diplomacy upheld by the United Nations that have been overshadowed of late by the politics of confrontation practiced by Trump and his global ilk.

1

How to COP

I first became curious about the UN-sponsored climate negotiations through my work in Bangladesh researching among people on remote silt islands (known as *chars*) north of the capital, whose lands and lives were in constant movement because of rains, wind, and floodwaters.[1] Over the course of my field-work, I noticed the trickle-down effect of global climate change–related discourse and programs. I even heard of an ordinary farmer who had traveled from the area to represent Bangladesh at the talks in Durban and afterward posted flyers around the *chars* inviting people to hear him speak about his experience. It was very unusual for *char* dwellers, who are among the poorest people in Bangladesh, to travel abroad, unless it was migration for work.

To learn more, I turned to Dr. Saleemul Haq, the Bangladeshi scientist–turned–international climate expert, who had visited my field site as part of a project scoping out places where community adaptation to climate change was taking place. His intent was to showcase local and national efforts at preparing for the looming reality. And he also attended the annual climate summits as an expert, aiding negotiators from the developing world. It was Saleem who drew me into the international aspects of climate change policy through accounts of his travel, descriptions of the many "hats" that he wore, and networks that he helped create. His words gave depth to the flat picture I had held of climate policy as operating exclusively top down. It felt like an extension of my fieldwork to see him in action. He was my conduit to the first meeting I attended in Paris in 2015.

Through my repeat attendance at the Conference of Parties over a five-year span (COP21 through COP25), exploration of meeting spaces, and growing understanding of the climate negotiation process, I came to appreciate the vastness of the effort to create a self-contained space, stretching across its many sessions, separate and free of the politics within host countries. As I explore in this chapter, tensions and contradictions within the overarching process manifested within the meeting spaces, just as the local context also often broke in. Climate activists, whom I located at the margins of COPs, were a constant reminder of the wider political stakes of the process.

In Paris, I found myself alongside 28,000 other attendees (unofficial count: 50,000) at the twenty-first Conference of Parties (COP) to the UNFCCC, a massive gathering being held in temporary, particleboard buildings in a suburb of the city. "Parties" refers to the countries that signed the convention in 1992 to take collective action to combat anthropogenic, global climate change. Every year since, the Parties had been meeting to negotiate binding climate policy.

Lost in the sea of words, I sought footing in the discourses around me and found it in a speech given by Asad Rehman. When I heard Asad for the first time—or, more accurately, found myself captured by his charismatic presence—he was head of the International Climate cell at Friends of the Earth International (FOEI), an international network of environmental organizations.

I asked Saleem about Asad. They knew each other from their years with the Climate Action Network (CAN), an umbrella group of ENGOs (environmental NGOs) that represented this constituency as a "non-Party stakeholder" within the negotiation process. Although happy to introduce us, Saleem warned, "Friends of the Earth International [FOEI] fought with the rest of the CAN members and left the group. They are very radical. They are never happy with any of the decisions at the COPs."[2] Given that I had just spent a week and a half in Paris bumbling around trying to comprehend some aspect of the busyness around me, and to develop stakes in the negotiation process but failing miserably, I was delighted by the possibility of meeting a naysayer. There were plenty within the academic world from which I hailed, but they were disparaging of the COPs because, although they wanted representative organizations and processes, they were repelled by the UN's fussy, procedural nature (see Jamieson 2014). A naysayer from within the process was a far more interesting prospect.

When I arrived in Paris before the start of COP21, it was only a few weeks after the bombings and shootings in Paris and nearby suburbs in which 130 people had died and for which the Islamic State had claimed responsibility. The country was under a state of emergency, and the annual climate march that had come to be associated with the COPs, where civil society made its presence visible and placed moral pressure upon the Parties and non-Party stakeholders, had been cancelled. The Saturday before the meeting was to officially begin, my colleague and I, who had traveled there to attend as members of civil society, joined thousands in a human chain organized by the environmental group 350.org in lieu of the march. Shoes were placed on streets to symbolize our marching. It was a poignant scene. I even caught a glimpse of Bruno Latour, the famous French social theorist, in the chain.

Amid the placard-holding crowd were several people with their faces elaborately painted; others wore masks and costumes. They entertained those gathered around them with pantomime and music. Afterward, we would again see these individuals on television when they resorted to forbidden marching and threw objects pulled from roadside memorials to the terror victims at the riot police, who in return tear-gassed and arrested them. My Hopkins colleague sighed and said, "It's all so French." 350.org would later distance itself from the violence, which was in protest of polluting factories and market consumerism.

By the time the climate demonstration descended into violence, I was already at the site of COP21, located in Le Bourget, a working-class suburb of Paris. I navigated long lines, had a picture taken of myself, got an ID, and went through security before finding myself in the official meeting space, the Blue Zone. The space consisted of enormous, convention center–sized temporary buildings with wide and long covered pathways running between them. One building was the designated site for civil society. Here booths were set up for different organizations presenting their environmental work and for environmental activists. The organizations ranged from the kind I have already introduced (FOEI, CAN, 350.org) to natural gas companies showcasing how their product was the clean alternative to coal and oil. Interspersed among the booths were small roundtables with chairs for use by the many civil society participants.

There were cafes and coffee shops liberally placed around the space, leading my Hopkins colleague to again sigh, "It's all so French." Certainly, the atmosphere felt like a first-class airport lounge, with the largely young, largely female volunteer staff all smartly turned out in black suits, with bright green or blue belts, walking around on tall stilettos guiding people.

The relatively open, central space of the building was encircled by rooms where panels, called "side events," were held each day, according to UN-identified themes such as "women," "farmers," "oceans," "youth," and so on. This was perhaps the most familiar part of the setup for me, reminding me of conferences I had attended in the past. The daily program was published early each morning on the UNFCCC website, available on the UNFCCC app, and broadcast over the closed-circuit television screens placed at different intervals in the Blue Zone.

The panels were further delineated by whether they were oriented toward *mitigation* or *adaptation,* two key buzzwords of climate policy. Mitigation involved efforts at scaling back existing carbon emissions, such as by making industrial processes more energy efficient or by replacing existing "dirty" sources of energy either with "clean" alternatives, so coal by solar, or through what the political commentator Naomi Klein (2015) calls "soft market options," such as carbon emissions trading, in which countries with low emissions sell their rights to emit to countries with higher emissions. Mitigation also involved more fantastical engineering schemes, ranging from making the upper atmosphere reflect the sun's rays to lessen its heating of Earth's atmosphere to capturing and sequestering the existing carbon in the atmosphere, which had already exceeded 350 ppm, or parts per million particles of air (see Hamilton 2013). In 2015, the carbon dioxide concentration in the atmosphere was already 410 ppm, which was enough for alarmed scientists to declare that we should not expect to scale back the effects of climate change in the twenty-first century and that we ought to be acting for the centuries beyond.

Adaptation, the second prong of global climate policy, was a latecomer to the negotiations and one that consistently received less attention than mitigation. It accepted the reality that climate change was already occurring (see Schipper 2006). It demanded attention to improving the lives of people considered most vulnerable to the suite of changes underway, so that they stood a chance (see Adger et al. 2006). In some ways, it was development by another name, but instead of thinking of the elimination of poverty as a good in and of itself, its elimination meant "greater resilience to combat climate change," to use policy language. The panels with titles like "From Science to Solutions: Uses and Strategies of IPCC Communications for a Climate-Changing World," "Provisions for Market Mechanisms in the 2015 Agreement," and "Investing in Resilience: Responding to the Adaptation Needs of the Most Vulnerable," provided a flavor of the presentations and discussions that were underway.

In the meeting spaces, I found myself in darkened rooms with stage lighting illuminating speakers whom I couldn't hear no matter how close I got. The UN had installed sound mufflers in the meeting rooms, possibly to keep down the din in the temporary meeting structures, but it produced a disorienting effect. You could speak to someone sitting next to you and still not be heard, nor could you hear the speakers on the stage. This muffling became more than an inconvenience, even producing a visceral sense of repression, when members of the various civil society groups, such as CAN, met informally in the rooms to discuss strategy and, even when sitting side by side, could not hear one another unless the speaker shouted. Donning headphones tuned to the language of one's choice quickly mitigated this problem, at least when attempting to hear the speakers on stage. A tech booth at the back housed the translators for each panel, facilitating communication at a conference marked by unequal relations and discordant voices.

After attending a few panels, I came away convinced that they were just program fillers to occupy the many members of civil society who milled about the conference site and that they lacked any significant role to play in the actual negotiation process. Later, at my second and third COP sessions, I would change my mind, but I think that is because by then I was caught by and in the process, like the anthropologist Jeanne Favret-Saada (2015), who was caught by magic practiced in the rural countryside of France in the 1960s through immersion in its language.[3] But for now, I will describe my initial, more cynical view, to convey my feeling of gloomy pessimism after a week and half.

I couldn't help wondering who was representing people already suffering and what it said about this process that the choices on offer were for renewable energy, when many lacked basic access to any meaningful energy source in the first place. My plaint was framed by my prior fieldwork experience on a silt island in the middle of the Brahmaputra River, where there was little electricity, running water, health services, and schooling for those living there. What could climate policy give those islanders more than what they were already getting in the form of development projects? My despondency primed me for Asad's radical bravado, but first here I will complete my description of the lay of the land, taking seriously the anthropologist Sally Engels Merry's (2007) contention that space reproduces the tensions and contradictions within an overarching process.

Between the civil society building and the building for Party delegations and negotiators was the media space. It was a cordoned-off area where reporters and bloggers gathered and had privileged access to computers that streamed panel discussions, negotiations (at least those open to observers), interviews with visiting dignitaries, and press conferences. Here too were the press conference rooms, which were in constant use as group after group—Party officials, civil society members, or UN officials—reported on different dimensions of the session. The press conferences were more compelling spectacles than the side events or even the negotiations, as presenters expressed a range of emotions, from outrage to excitement, and made off-the-cuff comments.

Past the media space, in the next building over, I found myself transported back to my childhood, to family trips to industrial parks and commercial fairs in Dhaka, Bangladesh, where countries set up shop to showcase domestic goods and international exports. At COP21, in the hall for Party officials, many countries hosted pavilions that were exponentially larger than the booths for civil society over yonder, which I had now in my mind downgraded to the status of mere kiosks. These pavilions were glossy and luxurious, with blown-up images, gadgets, screens, sofas, and free handouts. Judging by the crowds, the US and India pavilions drew the most attention in 2015. The US pavilion had a stream of high-quality, scientific programming and important figures on site to give talks and field questions. This image of a didactic, science-oriented approach to climate change, which had such an air of confident commonsense about it, stuck in my memory because there was to be no US pavilion in later years, during the Trump administration, and the side events the US government did put on were flagrantly anti–climate change. For instance, at COP23 in Bonn in 2017, the US government hosted a discussion on coal as a clean energy source, which was vigorously protested by participants (Holden and Oroschakoff 2017).

The India booth was also memorable for its sumptuous visual displays, including fountains, and one had to shake oneself out of the reverie they induced to ask how they served the purpose of environmentalism. After a long history of intransigence and inaction, in 2015 the United States was the good guys in the commentary I heard among those I met, because President Barack Obama was determined that something positive would come out of this COP (Goodell 2015), whereas India's and China's obstreperousness was repeatedly raised as an obstacle to a positive outcome (Praful 2012; Parker and Karlsson 2018; Eckersley 2020).

A quick brief on what was at stake in COP21: The Kyoto Protocol, which emerged out of the initial 1992 Convention, required countries to reduce their carbon emissions by a designated amount. While crafted in 1997, it was only ratified in 2005, and its first commitment period ran from 2008 to 2012. The time limit for the protocol had been extended to 2020, but countries were slow to ratify it, with its second commitment period only going into effect and ending on 2020. The goal of the meeting in Paris was to produce a more encompassing agreement, one with the cooperation of every country. The last time such an agreement was attempted was in 2009 in Copenhagen, but COP15 ended in such bad feeling, specifically between the developed and developing counties, as to be considered a shameful failure (see Marsden 2011). That failure had haunted every COP since. The widespread feeling in Paris was that either there was to be a global agreement there or else the entire negotiation process would be considered to have failed. Within this context, a cooperative, even enthusiastic United States was a pleasant surprise, whereas a difficult China and India was business as usual because, as large countries with developing economies, they were seen to be resistant to any moves that would curb their development. This impression of the two countries was also to be altered by later COPs.

The COP consisted of several conferences occurring simultaneously in Paris. There was the overarching Conference of Parties to the Framework Convention on Climate Change (COP), which was meeting for the twenty-first time. It presided over the Meeting of the Parties to the Kyoto Protocol (CMP). The eleventh session of CMP limped along, with little importance given to it. Instead, all attention was piled on the simultaneous meeting of Parties (again, "Parties" means "countries") to draft a new agreement under the aegis of the Ad Hoc Working Group on the Durban Platform for Enhanced Action, or ADP. Besides these three ongoing streams of negotiations (COP, CMP, and ADP), there were the sessions of the permanent subsidiary bodies, the Subsidiary Body for Scientific and Technological Advice (SBSTA), which ensured that state-of-the-art science fed into the negotiations process, and the Subsidiary Body for Implementation (SBI), which provided technical support on how the policies were to be put into practice, each on their forty-third iteration. The numbers assigned to sessions, meetings, decisions, articles, agenda items, among others, accorded with UN protocols and were second in importance only to their acronyms.

Five simultaneous conferences in one session took the notion of conferencing to an unprecedented scale. Large halls that could absorb thousands were earmarked for the major meetings of the bodies, such as the opening and closing plenaries. They were also open to civil society, for whom side rooms with a live feed of events were provided in case of overflow. Civil society was even invited to make statements within this plenary setting. Smaller rooms were set aside for the negotiations. They were open selectively; Parties retained the right to close off any negotiations to civil society. When that was the case, members of civil society waited outside to catch Party delegates with whom they might be acquainted, as they popped in and out of negotiations or ducked into their country offices. The delegates most often gathered in these offices, which were not public areas, although not cordoned off.

Delegates also attended side events and press conferences and granted interviews to the press in designated spaces. The building with the pavilions, delegate offices, plenary rooms, and meet-the-press sites was where the actual action happened, most notably the negotiations among Parties over policy texts for adoption. Finally, there was a third building with additional meeting rooms. Mostly intended for overflow, it had a neglected air. The walls were decorated with photo exhibitions both on the history of the negotiation process and scenes of climate change devastation already underway around the world. Alarming without being gutting was my impression of the tasteful images.

Art was clearly viewed as important, either to elicit further participation from civil society members, whose substantive contribution to the negotiations seemed at first minimal to me, or perhaps to distract them from the process. In Paris, giant, plastic sculptures of animals in primary colors in a Noah's Ark theme lined the pathways between the halls. It was more evocative of Disney World than the biblical story.

Where Is Asad?

As I walked around and discovered the zoned nature of openness and its parallel within the negotiations, with "informal consultations" and "informal informals" ensuring that most discussions were taking place beyond the live feed of negotiations being piped over CCTV, I started to have a sinking feeling that this was an orchestrated spectacle. Everyone here agreed that the issue of climate change was crisis-filled and imminent, but the procedural way it was being approached made the UN seem out of step with the times. It wasn't just a matter of a slow approach to a quickly changing situation. It was also that the structure of engagement was so embedded in old ways of thinking, so committed to putting humans—and only certain humans—before everything else, that it failed to acknowledge that the fabric of a planet in which both humans and nonhumans were interwoven was unraveling (Allen 2019).

I walked from the Blue Zone to the Green Zone several times. The Green Zone did not require the official badge that I used to access the Blue Zone, and thus civil society had a greater presence at these sessions. It was some distance to walk but still within the Le Bourget compound. This was by far the liveliest space, packed with people, including Indigenous peoples from around the world in full traditional regalia. They sang and danced not to entertain but to remind attendees of their experiences of marginality. There were booths, presentations, performances, and rallies going on constantly in this zone. Young people were everywhere, implicating adults for failing to protect them and ensure their future. Nevertheless, the space was haunted by a sense of disconnect to the negotiations. Speakers routinely and rhetorically asked, "Do they hear us?" referring to those in the official space. Despondent, I dragged myself to yet another side event: Fieldwork for an anthropologist is like that, one thing after another in the hope that a line of inquiry will click and some understanding will emerge beyond the clutter of first impressions.

And it did. I found Asad Rehman. On a panel entitled "Deal with It! People, Rights, Justice," I heard Asad, a British-Pakistani man, say that discussions over mitigation were edging out those on adaptation and finance in the negotiations. Local communities preferred to access funds directly from the Green

Climate Fund, a financial mechanism established under the rubric of the UNFCCC in 2010 to help developing countries tackle climate change.[4] I understood him as saying that a bottom-up approach to funding was more equitable and effective than funds funneled through governments of countries, known not only for corruption and mismanagement but also for privileging large infrastructural projects over more finely tuned, climate-focused interventions. Asad also warned that developed countries were currently double-counting the money that they were giving, as both development aid and adaptation finance. In other words, they were not making separate and additional payments into the funds for climate change. And he fretted that developed countries were trying to make those countries that were not historically responsible for climate change, that is, developing and least-developed countries, newly responsible for undertaking mitigation within the discussions in Paris.

His declarations, tinted with the conspiratorial, introduced a measure of spontaneity to the wider discourse in the COP, which had felt scripted to my ears. After the panel, I tried to catch up to Asad, but, speaking rapidly on his phone, he was quickly whizzed away for an interview with what seemed like members of the press. I left a note requesting to speak with him, but I would not hear from him at this COP. Instead, I only caught glimpses of him here and there as he participated in various protests at the margins of the meeting; his interviews were posted regularly on YouTube, however, which I watched.

I asked myself why I was so compelled by Asad and not by the hundreds of other well-intentioned individuals at the meeting speaking about the plight of this community or the importance of that neglected issue. I realized that I felt like an interloper at the COP, perhaps even a fraud, and Asad's words comforted me by suggesting that *everyone* there was a fraud. This could only mean that there was a more genuine reality, one that didn't feel quite so context-free, glossy, and full of products and optimism. If only to catch a glimpse of a different possibility, to see if environmental activism had a genuine alternative to offer, I made it a priority to seek out Asad at every session I attended.

The game of "Where is Asad?" became both a means to navigate the COPs and to learn about its margins, or, rather, to have those in its margins help me navigate the COPs. The anthropologist Laura Nader (1972) has called this way of acquiring knowledge "studying up." By this time, it had become clear to me that NGOs doing environmental work did not necessarily exist on the same spectrum as environmental activists. What environmental *organizations* and environmental *activists* meant by grassroots work was different. The nongovernmental organizations were more interested in programs associated with development with an environmental element, such as promoting self-sufficiency through generating one's own solar power. Activists were focused on changing the consciousness of people in the way, for instance, the educator Paolo Freire (2013) theorized, through acquiring greater knowledge of the workings of the global system.

COP21 concluded on December 12, 2015. Usually at the COPs the first week is given over to lead negotiators drafting text agreeable to all Parties, while the second week is the high-profile segment at which heads of state, ministers, and other high-ranking officials from around the world gather to give speeches and sign the negotiated text. At COP21, major leaders of the world gathered at the start of the session to inaugurate it, to mark its special-ness, as it were, and then a few gathered again in a lower-key fashion during the second week to sign the text and congratulate one another. The Paris Agreement was presented as the negotiated text to which all 195 countries gathered had agreed. In his concluding remarks, Laurent Fabius, the French foreign minister, who was also serving as the president of COP21 said, "The Paris Agreement allows each delegation and group of countries to go back home with their heads held high. Our collective effort is worth more than the sum of our individual effort. Our responsibility to history is immense" (UNFCCC 2015).

As Saleem had predicted, Friends of the Earth International declared the agreement a "sham" even before it was signed. Asad and others within his organization pointed to the fact that the agreement would not go into effect till 2020, five years on. So instead of attending to cutting emissions now, rich countries were effectively kicking the can down the road. In contrast, Saleem felt that here at last was an agreement, whereas before there had been none. Each position had its appeal.

In preparation for COP22 in Marrakech in 2016, I got busy trying to meet Asad. I was able to befriend him by means of Facebook. Asad responded to my efforts to message him by adding my name to a listserv for the Demand Cli-mate Justice (DCJ) movement, which I later learned had evolved from the Climate Justice Network, which had separated from CAN to serve as an alter-native organizing platform within the COP process (Keck and Sikkink 2000, Hadden 2015). Beyond doing me this favor, he avoided me at Marrakech, turning away from me the few times we came face to face. I understood the cold shoulder for what it was. Within the context of a high-level session, with pressure upon the relatively few activists present to be everywhere at once, digesting as many of the political developments within the negotiations as possible, serving as watchdogs and spokespersons, and protesting, when necessary, there really wasn't time to have conversations with an idle academic. And, besides, what had academics done for their movement?

RINGO, the constituency of Research and Independent NGOs, with which I was most associated at the COPs, seemed largely to be constituted of lawyers who were observing the process for research purposes. Like me, they were in awe of the scale of the process and curious to understand how everything was interconnected. In the standing language of the UN, they were interested in the "technical" side of the COPs and not the "political." At most, they aimed to advise Party delegations and NGOs.

For the environmental activists, in contrast, everything was necessarily political. Climate action, whether mitigation, adaptation, finance, technology transfer, or capacity building, had to be undertaken keeping in view not just the imminent nature of the global threats posed by climate change but also justice. Activists did not wish simply to see "justice" ensconced as a principle within the text of the agreement (see Forrester and Bell 2019). They insisted on "climate justice" (Martinez-Alier et al. 2016, Warlenius 2018). I understood this to mean that relevant Parties ought to take responsibility for creating the conditions for dangerous anthropogenic change and mitigate against it, while helping others develop the capacity to mitigate as well.[5] But it was also widely understood that in an unequal world, of which the anthropologist Ghassan Hage (2017) reminds us that the rule of force is the handmaiden of the rule of law, notions of justice were aspirational and ever deferred. The UN climate process had early on moved away from what it called the blame game to that of encouraging Parties to voluntarily take the lead in combating climate change. The very terms under which negotiations were taking place made it hard for charges of injustice or demands for justice to come up organically, requiring activists to make it their constant refrain.

Through the DCJ listserv to which Asad added my name, I became aware of the many different types of organizations and activists clustered around the COPs beyond those in CAN, whom I'd been introduced to the previous year in Paris. Although the COPs provided them a convenient place to meet each year and the DCJ a convenient umbrella under which to congregate, they were not equally invested in the UN or the climate negotiation process. Dismay at a global system that circumvented the issue of justice brought the activists together, but their critiques differed and kept them apart. The DCJ was a press of compelling voices continually threatening to become cacophonous.

But I get ahead of myself. COP22 was held in a convention center on a dusty plain in Bab Ighli on the outskirts of the beautiful, touristy city of Marrakech, which was bedecked for the meeting with exhibitions, musical performances, day trips to green facilities, and a general aura of anxious welcome. One day, on my way into the city to learn more about how they were showcasing climate issues there, my taxi driver flatly questioned why I was heading to the old, tourist part of town in the middle of the day when I ought to be with all my conference participants trying to come up with a solution to climate change. He noted that the delay in the seasonal rains was a source of concern for many Moroccans. It was 2016, and Morocco was experiencing more heat and less rain (almost 40 percent less than the year before) than ever before. And only some of the dry weather could be explained away by the regular effects of El Niño, the warm phase of the El Niño–Southern Oscillation, occurring every two to seven years.

At COP22, the usual bouquet of conferences was bolstered by two more, CMA1, inaugurating the meeting of Parties to the Paris Agreement now that the agreement had been ratified, and APA1-2, the Ad Hoc Working Group on the Paris Agreement, which was responsible for crafting the implementation guidelines of the Paris Agreement, which was being called the Paris Rulebook. I repeated the same steps I had taken at COP21, walking around the halls to take in the exhibits, booths, and pavilions. The art at COP22, also displayed along the pathways leading from one temporary meeting hall to another, was constructed out of scrap metal, wire, and fabric and ranged from tiny scarab beetles to foot-long, human-like figures busily navigating their surroundings. Searching them out was fun, as was gazing at them, for they seemed as full of energy as the officials rushing about around them. Figures of two youths holding hands, one with what appeared to be a solar panel for a head, and of a waste picker pushing a wheelbarrow subtly brought in a dimension of difficult, but scrappy, everyday life in what was otherwise a quite official-feeling space.

Once again, the days were organized thematically. I narrowed my list to specific topics so as not to get overwhelmed and to be able to follow developments more carefully. Given my primary field site of Bangladesh and mode of entry through Saleem, I had started at COP21 in Paris with a focus on adaptation. This focus had branched into finance, as directed by Asad, and into a new issue, that of loss and damage.

Loss and damage had been gaining ground for some time and, with its enshrinement as an article in the Paris Agreement, was being discussed at Marrakech (Surminski and Lopez 2015). With loss and damage there was finally acknowledgment within the negotiation process that climate change might not be mitigated and that there were limits to human and ecosystem adaptation (Adger et al. 2009). While some damage could be rolled back, there was going to be—in fact, there already was—irreparable loss. For developing countries, loss and damage meant the loss of lives and livelihoods. They agitated to make it a potentially legal means to put pressure on developed countries to mitigate or else pay out. For developed countries, it was a market opportunity to develop new insurance products involving risk sharing and transfer. There was no clear resolution in sight (Calliari, Serdeczny, and Vanhala 2020).

In addition to attending side events and visiting pavilions on the themes of adaptation, finance, and loss and damage, which was already quite a lot, I sat in on a few negotiations that were open to civil society. I noted that negotiations on loss and damage were routinely closed to civil society observers at the request of Parties, most likely because of the controversial nature of the issue. I attended press conferences with a range of actors—environmental activists from FOEI and CAN; national delegates from the United States and Bangladesh; representatives from transnational groups such as the Pan African Justice Alliance and from UN bodies such as the United Nations High Commissioner on Refugees and the World Bank—to understand how each was positioning itself, navigating this complex and crosshatched space, providing analysis of daily developments, and highlighting pathways in negotiations for their various audiences to follow.

I also started to attend the daily meetings of RINGO to hear from fellow scholars about what they were learning at different parts of the COP. The daily meetings, which I would come to rely upon more and more with each passing year, both for a broader understanding of the COPs and for tickets to negotiating rooms, were the crucial means by which I felt myself submit not only to the punishing schedule and pace of the conferences, but also to something like a community of interest. I felt the UN entraining us, essentially a group of drifters, into its workflow, as we tried to make ourselves useful by writing up our notes on the negotiations we attended for our fellow researchers, in exchange for the dubious privilege of attending the negotiations. I also started to read the daily briefing papers put out by non-Party stakeholders, *ECO* by CAN (in publication since 1972), *ENB* by IISD, and *TWN* by Third World Network, of which the last was the most left-leaning.

And, most importantly, for the game I had started with myself at the Paris conference, I sought out Asad. While he avoided me, he was at the COP to be heard, speaking in his rapid-fire pace at Friends of the Earth press conferences, meetings of the DCJ, and in YouTube interviews. Even as I began to identify phrases in his speech as well-rehearsed soundbites echoed by other left-leaning activists, positions that were as well crafted as those of any other official within the UN process, there was still something in his speech that tugged at me. It was his ability to be completely immersed in the moment and, simultaneously, see the process from the outside. Drawing on ancient Greek rhetoric, the philosopher Michel Foucault (2001) has identified this mode of speaking candidly in the face of power, for the common good and at personal risk, as *parrhesia*. Even as I oriented myself within the COPs through a focus on issues, I wondered if I wouldn't have been equally well served by following the passion of parrhesia to gauge what issues mattered most.

In Marrakech, Asad spoke of finance, or the lack thereof, pretty much as he had done in Paris. While $100 billion per year had been arbitrarily decided upon as the amount that developed countries were to put annually into the Green Climate Fund to help developing countries combat climate change, no one was yet speaking realistically about the inevitable costs of transformation, he rumbled. The more realistic dollar amount needed was closer to $4 trillion, he claimed. Regardless, thus far only a fraction of the $100 billion/year had been pledged and an even smaller amount placed in the coffers of the fund. Nor were the developed countries doing as much as they should to transform their economies and societies to keep with the pre-2020 emission goals set by the Kyoto Protocol, as affirmed by the 2012 Doha Amendment. And, as before, he brought up "historical responsibility" as an aspect of climate justice.[6] The industrialized countries that historically contributed the most to climate change now bore the lion's share of the responsibility to mitigate. Developed countries were failing to meet this obligation, despite efforts to remind them by social movements, environmental and developmental NGOs, trade unions, and faith-based and other civil society organizations.

COP22 was a transitional meeting. The Paris Agreement had been ratified just a short while ago. It was understood that the Party delegates were tired and wanted to take it easy. They had no intention of formulating any major policies in Marrakesh, leaving the work of producing a rulebook to the COPs to follow until 2020, when the agreement would go into effect. The more relaxed feel at COP22 gave me the opportunity to take in details I had missed in Paris. For instance, there was a pop-up called the Climate Action Studio, later renamed Action Lab, where climate change action around the world was showcased throughout the day, in forty-five-minute segments. Unlike the side events and press conferences that were attuned to daily developments within the negotiations, the studio was more pedagogical, aiming to represent organizations and best practices. I heard US-based scholars talk about how they were experimenting with media-style communication on climate change to generate urgency among members of the US population, Green Climate Fund presentations on how to apply for funds, explorations on how Indigenous knowledge might be mobilized within this process, among other talks archived and made available through YouTube. Given that these topics would go on to be significant in future sessions—for instance, Indigenous knowledge would become a platform in its own right in 2017—it suggested the importance of following the studio to gauge future trends within UN policy making.

Activism at the Margins

The daily program, published and broadcast early each morning, was a very complex artifact, opening up to further and further events and programs only if one knew to follow innocent-looking links and keys. Most astonishingly, it held a schedule of daily rallies and protests within the two zones, politely titled, "List of Authorized Publicity Actions." These included CAN members passing out copies of the *ECO* flyer summarizing the state of the deliberations to participants entering the Blue Zone each morning and its famous Fossil of the Day gathering at the end of each day. At this latter event, CAN members dressed up in diverse costumes, delivered satirical speeches, and gave out awards to the country or countries that had been most obstructive within the negotiations that day. It was advertised as: "Find out what countries have done their 'best' to block progress in the negotiations. Real stage, real awards, real serious."

Other actions advertised included one-off rallies such as one by Maori youth groups from New Zealand called "Who Wants to Be a Youth Leader?" This was described as "a mock game show where youth from countries such as New Zealand, Brazil, France and the UK are asked whether their country was being ambitious in tackling climate change. They hold up signs with their country's response either saying 'Yeah', 'Nah' or 'Yeah, nah'. This will be followed by chants calling for these countries to have more ambition, and speeches by youth representatives from them." From this program it first became clear to me that individuals involved in the actual negotiations *and* those critical of them were within the UN process, requiring acknowledgment, credentials, and permission to carry out their work.

Even offsite events were often listed in the daily program, such as the climate march, which was historically held during the weekend between the two weeks of the session. While the climate march in Paris had been cancelled the previous year, the march in Marrakech was on and well advertised. On November 13, the Sunday that fell in the middle of the two-week meeting, thousands of people gathered out front of the stadium in Marrakech to demand immediate and urgent action by the participants at COP22. What gave their demands an added urgency was that Donald Trump had won the presidential election in the United States on November 8, and one of his campaign promises was to withdraw the United States from the Paris Agreement.

With the United States poised to pull out, the future of the agreement was uncertain. Everyone wondered if the remaining developed countries had the will or the capacity to provide the necessary leadership to go forward with adopting climate change policies without US participation. Given that the United States was one of the biggest polluters in the world, if it continued to pollute while other developed countries incurred the expense of reducing their pollution, it stood to give the United States an unfair advantage in the global economy, what is called the free-rider problem. It wasn't even clear if the combined actions of the remaining developed countries could make up for the deficit in emissions reduction that would be brought on by a possible US withdrawal. Both scenarios meant that developed countries were inclined to stall on making any progress in the negotiations until and unless it became clear what the United States was going to do. And what did the US withdrawal mean for large developing countries such as China, Brazil, and India? Was it an opportunity for them to show leadership under the new paradigm in the climate change regime, which had moved away from blaming and shaming in the mid-1990s to encouraging leadership in combating climate change? Or would they take their lead from developed countries?

With so much up in the air, it was no wonder that the gathering at the march was large. There were public announcements by organizers and participants that their intention was to broadcast to the world that climate change remained an important issue regardless of geopolitics. There were gut-wrenching placards and posters in view. People came with props but without the facemasks and face painting seen in Paris. It took a while for people to gather, but, once assembled, their energy was palpable. The crowd started to march toward and then down a main boulevard of the city.

Then something happened that showed me that, although disallowed from entering the actual spaces of the sessions, the public marches organized during the COPs were important sites for the expression of local politics. Just as anarchist provocations interrupted the human chain in Paris, pushing against the limits placed by the state of emergency, so did fault lines within Moroccan activist groups in their opposition to the monarchy express themselves at this march. When the protesters reached the boulevard, one contingent of marchers turned around and headed back to the stadium, while the rest continued down the boulevard to the endpoint designated for the final rally. I went one way, then the other, and then backtracked to follow the first group, confused and lacking the political acumen and linguistic skills to figure out what had just happened.

There was very little media coverage of this split in the march, which was internationally heralded as an impressive gathering of protesters. But from what I was able to gather from the website of Attac Maroc, the Moroccan branch of an international organization that described itself as part of an alter-globalization social movement and that had spearheaded the anarchist action in Paris, lo-

cal NGOs had had a falling out over whether to organize a march or to boy-
cott COP22 entirely, because having it in Morocco served to greenwash an
authoritarian government whose actions, undertaken in the name of climate
change, were deleterious to human rights. The reference here was to the longstanding denial of self-determination to the Saharawi, the Berber-Arab inhabitants of the Western Sahara, where the Moroccan government had most
recently constructed a large solar power plant on disputed territory. The dissident activists instead sought to throw in their support for the Berber inhabitants of Imidir, a small mining town in the Atlas Mountains, which had been
in a protracted standoff since 2011 with an international mining company to
prevent its waste from polluting Imidir's waters. While the decision was made
by Moroccan activists to rally, the subsequent split in the march indicated dis-
avowal by dissident activists of the major Moroccan organizations, which
were considered complicit with the government.

It was only the following year at COP23, in Bonn, Germany, in 2017, when
I had finally eased into a conversational relationship with Asad, that I found
out about the contradictions global environmental activists faced in putting
solidarity into practice. I learned that activism was very restricted under the
Moroccan constitutional monarchy. Moroccan activists had to negotiate
heavily with the government to be able to host and organize the international
march for the climate that had come to be associated with the COPs. International organizations, such as Friends of the Earth International, felt duty bound
to participate in the march to show solidarity with local activists, as well as to
keep the pressure on the COPs. At the same time, such acts of solidarity
came at the expense of full-throated support for those affected by Moroccan
state policy and climate action, making DCJ's claims to speak for climate
justice ambiguous. Not everyone in the DCJ network was comfortable with
this position, and the disgruntlement spilled into the WhatsApp exchanges to
which I was privy.

Back in Marrakech, a hint of the challenges of activism was evident in the
anguished soul searching of Nathan Thanki, a pensive young man of Scot-
tish and Gujarati descent with a long history of involvement in the COPs and
a close associate of Asad. As the convener of the DCJ, it was his intensive maintenance of the listserv during the COPs that held the DCJ together in discussion and coordinated action. One evening, Nathan and I swiped drinks off the
refreshment table at a reception and talked about COPs past and present.
While it was clear that Nathan understood the process inside and out, having
started as a youth member, he despaired as to whether it was any longer possible for activists to make meaningful contributions within it. "At one time activists used to stand over the heads of the negotiators heckling them to do more

and move faster. Now we have to watch negotiations from overflow rooms and read out pre-prepared statements in the minute or minute and half granted [each constituency]. In one way, it's better that observers have been parsed [referring to their groupings into constituencies, such as environment, business, women, research, trade unions, youth, etc.] so no one group can co-opt the voice of civil society, but the activist space has been steadily shrinking," he bemoaned. Nathan wondered whether activists such as himself kept returning simply because they were addicted to the process, exhilarated by their ability to keep pace with its growing complexity and serve as interlocutors to outsiders.

It was beginning to dawn on me that although Asad's speech had a spontaneous quality and attempted to speak truth to power, activists at the COPs were called upon just as much as others to play their part, and their parts were equally scripted. In fact, they risked becoming performers at the sidelines of the COPs, with delegates stopping to snap pictures of protesters in their colorful masks and costumes before continuing to official sites cordoned off from civil society.

Equity by Other Names

I experienced COP23 in Bonn in 2017 very differently, more optimistically, than the previous two COPs because my angle of entry was different. Getting the go-ahead to be present during the DCJ's two-day assembly before the start of the meeting in Bonn meant that I was immediately immersed in wide-ranging discussions about the challenges faced by activists around the world. Nathan led the assembly with Lidy Nacpil, a Filipina veteran activist who was the coordinator of the Asian People's Movement on Debt and Development. They insisted that although the COP provided the opportunity for many activists to come together, it should not dominate activist discussions. The focus was very much on country- or region-specific challenges, ranging from unfair labor conditions to exploitative resource extraction. At the same time as the activists spoke about local challenges, they also espoused an awareness of how their individual countries were representing themselves within the negotiation process, drawing attention to obvious gaps between rhetoric and practice.

Their presentations were impromptu, sometimes hyperdetailed, for instance, outlining the percentage of Germany's economy driven by coal or enumerating how many polluting corporations were involved in the negotiation process. At other times they offered more of an overview, for instance, discussing how climate developments in Africa were not yet focused on the needs of rural people. Everyone underlined the fact that the environmental aspect of climate change was only one face of the problem. The other aspects included authoritarian governments, exploitative corporations, compromised police and judiciary, strained resources, and so on. Climate wasn't only an environmental problem but a societal and political one. DCJ organizing appeared to be three-pronged, which was to strategize on how to help activists apply pressure at specific points within their countries, decide which major international meetings and gatherings of heads of state to attend to maximize global activist presence and pressure, and plan actions for the COP about to start. Clearly, the participants were strapped for people and resources. The constant refrain was that the activists commit to helping one another organize actions, attend meetings, participate in one another's actions, and, most importantly, ensure that their actions did not conflict schedule-wise. Just like in Marrakech, solidarity showed itself to be a vexed practice. At one point during the assembly, a young woman turned to me to ask pointedly why a particular activist was present at the assembly given the allegations of sexual harassment against him. I had no answer but could see that he was close to the organizers, providing important in-country actions and participating in COP-related actions.

The DCJ maintained a strong ethos of not centralizing or dictating organization, even if that resulted in somewhat dispersed and DIY activism. I understood this ethos to have developed in opposition to CAN, the other major network of environmental organizations, whose modus operandi was to enforce a single voice among its constituents to be most effective within the negotiations. The DCJ felt that this approach gave the negotiations the upper hand, forcing activists to take on the technical language of the process and to make compromises on pragmatic grounds. In contrast, DCJ wished to be unequivocally clear that they sought climate justice and stood in solidarity with all those marginalized in the process and beyond.[7] In *Networks in Contention: The Divisive Politics of Climate Change* (2015), Jennifer Hadden explores the history and dynamics of environmental activism within the COP negotiations, showing how and when the split occurred between those who sought to ensure that science dictated policy and those for whom justice was paramount. She writes how, for the former, any deal was better than no deal at all, whereas, for the latter, the political process was more important, with no deal preferable to a compromised process.[8]

Unlike previous years, I didn't attend COP23 until the third day. Instead, I spent the first two days following Lidy as she participated in the People's Summit, which was organized as an alternative COP for those seeking to influence the process from outside. Such spaces were at a greater physical and conceptual distance from the process than the activists operating within the Blue Zone or even those operating in the Green Zone. However, this distance did not imply that the alternative spaces were irrelevant to the process. Rather, it suggested the width of the margins encircling the COPs.

Lidy was as powerful a speaker and as charismatic as Asad. However, while Asad drew his power from clipped speech that suggested reserves of righteous anger, Lidy spoke more expansively, sometimes apologetically, but always morally. Her voice was like a conscience; it vexed more than it stirred. During one of the interviews she gave in the days that I shadowed her, she explained the thinking behind climate debt and the attendant notion of "fair share."

Developed countries of the world had achieved their current levels of economic success and standards of living by polluting and in the process had produced the problem of climate change. Furthermore, it was the pollution of developed countries that clogged the atmosphere, leaving little space for further pollution. Whatever space remained within the atmosphere ought to belong to developing countries. It was their fair share.[9] While developed countries tended to be creditors to whom poor countries were indebted, historical responsibility reversed that relationship. Now developed countries owed a debt to poor countries, if not to humanity or even Earth, for having brought about this problem. And they could only repay their debt by letting other countries pollute while they scaled back their own pollution. This was a problematic proposition, and my kneejerk reaction was to urge that all countries should stop polluting, that polluting was an absolute bad. But Lidy's words nettled me into entertaining the idea that in an unequal world in which developing nations had been historically held back and in which getting ahead meant polluting, whatever scope remained for polluting ought to be the preserve of those hitherto marginalized, if justice was to be served.

COP23 was a Fijian COP, though it was taking place in Bonn. It was the first time in the history of the COPs that an island nation was organizing the annual meeting. However, since Fiji could not accommodate such a large influx of people, the German city had stepped up to host the event for them. Bonn was also the site of the UNFCCC central headquarters, the UN's university, and the former parliament building of the West German government, which had been transformed into a convention center after the capital of reunified Germany moved to Berlin. The negotiations were held in a composite of the three spaces, but since the buildings could not hold all the participants and projected activity, a second site was erected in the middle of a park in Bonn 1.2 kilometers from the main meeting space.

After the COPs in Paris and Marrakech, where the Blue and Green Zones were within walking distance of each other and their activities interlinked, the location of the space for civil society organizations felt overly far from that of the official negotiations. Activists fretted that it would have a dampening effect on civil society participation within the negotiation process. They were not wrong. The negotiations, along with delegate offices and press conference rooms, were entirely contained within what was called the Bula Zone, in a nod to the Fijian Presidency, whereas all the side events, booths, pavilions, exhibitions, and performances took place within the Bonn Zone. COP23 was effectively two separate events, one marked by the official protocol of negotiations and the other by a festival atmosphere in which Fijian culture was bowdlerized for the purpose of making it consumable for an international conference. A Talanoa space for the respectful exchange of ideas was set up in the Fijian Pavilion, decorated with traditional furniture and cardboard cutouts of picturesque places. It was in constant use the two weeks COP23 was in session. Whereas activist actions within the Bula Zone felt weak and isolated, those in the Bonn Zone were febrile and heady, but there were no clear lines of influence and communication between the two.

The 1.2-kilometer walking path between the two zones showcased a wide palate of climate change–related art that aspired to monumentality, in contrast to the diminutive but emotionally moving metal sculptures in Marrakech. There was a life-size teepee made from squares crocheted by climate-affected peoples and a large metal sculpture of a polar bear impaled on a spike. The German Forest Association built a tree with wooden planks over the course of the two-week meeting, but their message seemed to be no more provocative than to remark on the importance of trees for humans, or perhaps to show that trees cannot be reassembled from timber. There was a sixty-five-foot giant inflatable Earth Globe balloon sponsored by the German Development Ministry. Perhaps the most prominent and visited site between the Bula and Bonn Zones was the unofficial, alternative US pavilion, called the US Climate Action Center. It bore the logo "We Are Still In," referencing Trump's initiation of the legal process of pulling the United States out of the Paris Agreement.[10]

By this time, I had decided to write my book on the COPs, with a focus on locating Bangladesh within the sessions' many playing fields. In addition to following Saleem, I had started to pay attention to the official Bangladeshi delegation, the experts guiding them as part of the least-developed-countries constituency, and the activists representing Bangladesh as part of the larger activist scene. But I was still issue driven in my navigation of the space and schedule. My issues remained adaptation, finance, and loss and damage, although these had separated out into many strands. For instance, adaptation was now spoken of both in terms of *adaptation communication*, that is, whether Parties should or shouldn't communicate their adaptation efforts within their

Intended Nationally Determined Contributions to carbon mitigation for sub-mission to the UNFCCC as part of the Paris Agreement, as well as their ef-forts at educating their own citizenry about climate change science and policy, and *adaptation finance*, as in who or what was going to fund developing Par-ties' transition to climate-friendly economies and lifestyles. Finance had gone from being a single issue (how much developed countries should contribute to the Green Climate Fund) to being an aspect of every issue, including ca-pacity building and technology transfer. And loss and damage was in peril of becoming a marooned issue with no home in the conversations taking place simultaneously across the various climate conferences.

Simply trying to understand how each issue was woven into the skein of negotiations became a Herculean task. Once again, I attended the daily RINGO meetings, where I could track negotiations through other members' reports. I went to a few press conferences, but now I more tightly focused on the constituencies I was following to see how they were positioned. Similarly, I went to a few side events and kept an eye on the schedule of the Climate Change Studio/Action Hub. And I noticed that the daily program offered more than I had been able to grasp in previous years when I was still getting my bearings. For instance, the Secretariat of the UNFCCC regularly sponsored pedagogical events aimed at educating attendees on the growing number of technical issues associated with the Paris Agreement. There were events man-dated by previous COP decisions. There were also high-level events to signal partnerships between Parties seeking to undertake climate change on their own terms outside of the negotiation process but who were using this platform for enhanced publicity.

I also noted with interest intensified activity at both the China and India pavilions. Everyone continued to wonder aloud what the two countries were going to do now that Trump had effectively pulled the United States out of the deliberations. Were they going to take on "leadership," in the parlance of the COPs (Urpelainen and de Graaf 2018)? It was said that Europe was hiding behind US inaction. China and India, in turn, talked about how the devel-oped countries had yet to meet their pre-2020 commitments or discuss what finance they were going to put on the table to help other countries mitigate and adapt. And they complained that developed countries were driving a wedge between countries by deepening the distinction between developing countries and least developed countries.

Asad didn't participate in COP23 at Bonn. By this time, he had left Friends of the Earth International and joined the War against Want, organizing against poverty and injustice. I caught sight of him at the pre-COP DCJ Assembly. He was there to strategize with other founding members of the DCJ, Nathan and Lidy among them. He finally had some free time and invited me to join him and his friends for dinner. On the tram en route to the restaurant, Asad shared his backstory on how he became involved in climate change. He had begun life as a scrappy Pakistani boy in Manchester, dealing with the racism that daily life in the United Kingdom threw as a matter of course at its migrant communities. His childhood experiences led him to become a community organizer, making alliances with a heterogeneous crowd, including punks and religious leaders of mosques.

Through his activism, he ended up at Friends of the Earth. His decision to move into the climate change cell within the organization surprised many, because the issue seemed remote from the lives of migrants and tied to elite institutions and scientific communities. However, Asad said he wagered that global climate change policy could learn from experiences on the ground. Also, advocating for fairer and more just climate policies from the get-go would help avert future top-down action. Climate policy was ultimately less about the environment and more about justice for him. At dinner I noticed that he introduced me as a comrade from Bangladesh and didn't make any mention of my being a US-based academic, suggesting the uncertain location of academics within environmental activism.

I saw Asad one last time at the twilight of COP23, when he came for a meeting with the gathered DCJ leaders. We chatted; this is when he explained the dynamics of the split march in Marrakech the previous year, pointing out that activism in each country brought with it its own unique challenges. All international activists could do was support activism in principle, without being judgmental or getting pulled into country-specific dynamics. Case in point was what happened with Ende Gelande. The German activist group called for occupying lignite coal mines near the COP23 venue on the eve of the meeting to highlight the hypocrisy of the German government, which positioned itself at the cutting edge of clean energy all the while mining the lowest-quality coal. Yet occupying an active coalmine suggested deafness to the cause of "just transition" important to trade unions and others representing workers within the negotiating process. DCJ, however, would not adjudicate on this contradiction and joined the protest. Despite the clash between Ende Gelaende activists and the state, the climate march at Bonn went off as planned and had a large turnout, in comparison to the marches in Paris and Marrakech. It suggested a fairly unified environmental movement within Germany, one that was largely in sync with its government despite differences over Germany's continued extraction of dirty coal.

At our last meeting, Asad and I sat among the many other civil society observers who waited for the conclusion of the high-level portion of week two of

COP23, that is, for negotiations to end and for the final decisions to be made public. This wait can carry into the wee hours of the last day of the meeting, into the next day, and sometimes through the weekend. Bonn was the first time that I participated in this waiting period. While those gathered explained that such last-minute extensions of the meeting, deferrals on decisions, and even disappointing outcomes were a part of the ritual of the COPs, there was excitement in the air. The possibility of commitment to a final declaration, one that would have a ring of gravitas appropriate to such a large and costly gathering and that carried global significance, made these last hours suspenseful. The moment might produce nothing but an agreement to continue negotiating, as in Copenhagen in 2009. On the other hand, it could entail the rare capitulation by developed countries to the demands of least developed countries and small island nations, such as happened in Warsaw in 2013, leading to the creation of a mechanism for loss and damage within the process. The possibility of such a victory, however tiny, made the wait worth it.

COP23 in Bonn in 2017 was an important one, both because of the historic nature of its presidency and because decisive progress had to be made if the Paris Rulebook was to be finalized by 2018, so that it could be implemented by 2020. What I followed of the negotiations thus far suggested to me that the rulebook was less a meaningful policy about climate change and more of an indication of buy-in from every Party into the reality of climate change. It was about instituting a global system of monitoring to ensure that every Party was including climate considerations within its economy. One can ask whether this was the meaningful change people in the world were looking for. But the more noteworthy fact was that this was clearly a process to which every Party seemed committed, despite the split between developed and developing countries and, as I was to find out, the split within developing countries.

The Local Context Breaks In

Between Bonn in 2017 and COP24 in Katowice, Poland, in 2018, there were two further meetings to ensure progress on the Paris Rulebook. The Polish Presidency promised that it would deliver on the rules for implementation at its COP, and it did, for the most part. It was at Katowice that I became convinced that national contexts and local politics shaped UNFCCC outcomes, despite all efforts to make the climate process appear standardized and self-contained. I may only be stating the obvious, because for all its emphasis on procedure and technical work, the UNFCCC process was political through and through and therefore needed something like political will to propel it. But I think it is important to show whose will came to matter and how.

This was Poland's third time hosting a COP, and none of my activist inter-locutors were pleased about it. There had been a definite shift to the right in the country in recent years. Although eager to serve as host, it had a poor rec-ord of climate action and a deep suspicion of environmental activism. And there was widespread fear that the country would use the occasion of COP24 to usher in draconian surveillance and security measures, which would be det-rimental for organizing at the conference by activists coming from abroad, not to mention for the domestic civil society constituencies. This fear was realized.

I thought I had seen something of the impact of state repression on envi-ronmental politics in Marrakech, but Poland put those efforts to shame. Firstly, COP24 was in Katowice, a major city in coal country in Poland.[11] Pride in being a mineral-rich region found expression in a large, impressive, and bru-talist metal sculpture that commemorated patriots in a park immediately in front of Spodek, the spaceship-styled convention center where COP24 was held.

Coal was the lifeline of the region, and chunks of it were sold as souvenirs all over the city. To hold a conference on climate change in such a place was to rub everyone's face in the fact that not only was Poland coal dependent but also proud of it. Hip young Katowice city dwellers and families at the Christ-mas Market in the city center with whom I chatted laughed ironically when I told them that I was here for COP24. They asked if I found the air difficult to breathe. They had been told that air quality was terrible in Katowice, but they had no point of comparison. And they offered me their cigarettes: What dif-ference could it make in this place?

In addition to putting a new law into effect enabling the government to bet-ter monitor activists around the COP and holding the conference smack in the middle of coal country, the housing for the attendees was so far from the meeting site that it was virtually impossible to walk, bike, or take public trans-portation there. Instead, attendees had to rely on Ubers and taxis. The Polish government that held the COP presidency had chosen perverse ways to show its commitment to global climate action. Such initiatives were expected of every government that hosted the COP, which also served as an opportunity for governments to highlight domestic or regional climate activities on host-ing a major international conference.

Among Poland's initiatives was one titled the Just Transition Declaration. On the face of it, it sounded very positive. It was to highlight the plight of workers, particularly those in emission-intensive industries, as economies transitioned to low-carbon pathways, and to ensure that they were also part of the transition. Just like other civil society constituencies, such as research organizations, environmental groups, youth, etc., trade unions had their own constituency, called TUNGO, or trade union NGOs. I had early on learned that their mantra was "just transition" and that the widely used phrase within activist circles "climate justice" sought to include workers affected by mitigation efforts among those affected by actual climate change.

It had been a broad coalition of civil society constituencies in Paris that had ensured that all vulnerable and affected groups would find mention in the Paris Agreement, even if largely symbolically in its Preamble. By highlighting only workers and by promising to protect them in the middle of a coal-dependent region without any evident plans for transitioning out of coal, the Polish government was essentially extending support to coal.

Tetet Nera-Lauron, a Filipina activist friend who worked for the Rosa Luxemburg-Shiftung on climate politics, told me that TUNGO's decision to support the just-transition initiative of the Polish Presidency came at the expense of the broad coalition that had been forged in Paris and that had continued working together since to ensure not only that climate mitigation was taking shape but that there were protections built into such activities so that vulnerable people would be protected both from the impacts of climate change and from the actions taken to combat it. She feared that, legitimized by the support of workers and trade unions, the Polish Presidency would not pay any further heed to the concerns and demands of other civil society constituencies. This fear too came to pass.

The climate march was held the first Saturday of the COP on December 8, 2018, in the midst of these tensions and fractures among the activists. TUNGO did not participate in the march, although many other organizations, NGOs, and individuals were there in full fury, gathering, chanting, marching, and speechifying. While it was an incredibly energetic event, it was overshadowed by the fact that police in combat gear lined the sides of the street in such quantities as to almost outnumber the marchers. While we were used to plentiful security guards around COP sites, this police presence felt oppressive, even heavier than in Paris in 2015, in which there was ultimately a faceoff between the police and anarchists. The difference was that while in Paris the temporary imposition of martial law aggrieved the climate activists and others, in Katowice the police presence felt more like business as usual. Some twelve activists, including Zanna Vanrenterghem, of Climate Action Network Europe (CAN-Europe), had been stopped at borders, either turned away or detained to be released later.

Such actions were intimidation tactics, powered by the bill that the Polish government had passed earlier in the year giving it broad authority in making arrangements for the COP. However, it was faced by the tactics of shame and condemnation best expressed by the fifteen-year-old Swede, Greta Thunberg, who gave her first ever speech at a UN climate meeting in Katowice on December 12, 2018. Her presence signaled the emergence of the youth as a global constituency as important as any other.

While I attended COP25 in Madrid in 2019, I'll forgo describing it until the conclusion, as it serves as a natural way to draw to a close my observations of the COPs initiated in this chapter and unfurled over six more. Suffice to say that I was perhaps at my most aware and attuned in Madrid, since it was the sixth session I attended. But I was still not entirely prepared for the depth of despair expressed by activists at the presence of the very corporations aggravating the problems of climate change. They were at COP25 as part of the private sector, where hopes were increasingly being placed for mitigating climate change. The session ran over by forty hours, a hugely costly and unprecedented delay, and ended in a hung jury, with almost all deliberations on major issues struck from the record, since they had not resulted in decisions. This failure, as declared by the media, occurred despite a palpable sense of impending crisis personified by banners and posters with chilling messages like "Tick tock. Tick tock. Don't call it change. Call it climate emergency."

After I returned home, I cooked for two days, then hosted a dinner for family and friends in anticipation of the end of days.

2

The Voice of Bangladesh

I had come to the COPs to study how Bangladesh fared within the climate negotiation process. I did not approach the delegates immediately, preferring first to see how they presented themselves in public. Initially, I focused largely on Saleem, his team, and their extended network. I interviewed Bangladeshi activists who had thrown in their lot in with the Demand Climate Justice movement. I met members of the Bangladeshi press corps. In addition, I encountered a number of prominent Bangladeshis entrenched in the various UN agencies that were part of the process, as well as expatriate Bangladeshi youth who had come to the COPs for training or internships.

I thought that this level of presence for a nation was standard until someone commented to me that Bangladeshis seemed to be very well represented at the COPs. It tallied with a perception I had from prior work with the United Nations High Commissioner of Refugees in Bangladesh, that for educated Bangladeshis, UN-led processes provided opportunities for employment, out-migration, and upward mobility. At the same time, they engaged in these opportunities as nationals, that is, proudly asserting their national identity and desire to up build Bangladesh's reputation and standing.

So it came as some surprise to me to encounter two scientists within the Bangladeshi delegation who spoke distrustfully about the process and about Bangladeshi participation in it. Drs. Ainun Nishat and Atiq Rahman, both well-known scholars of the environment and climate in Bangladesh and abroad, sat down with me one day at COP23 in Bonn and pronounced that while attending the COPs was a lark for some and an addiction for others, there was no further use for the process for Bangladesh. The country already had its future tied up with the World Bank, China, and India, and there was nothing it could or would do to change course in terms of sustainable development, much less carbon emission reduction. Bangladesh's participation in this process was mere lip service. Dr. Nishat ventured that I was far gone in my addiction to the COPs, as evidenced by my returning year after year, a seduction of which the activist Nathan Thanki had also earlier warned me.

When I asked Saleem about Bangladesh's place within the COPs, his answer was somewhat more encouraging. He said that although Bangladesh's fate may already be determined because it was, after all, a small, poor, dependent country, it was admired for taking a leadership role at the COPs and for inspiring others. In this chapter, I draw out what it meant to be a small, poor, and dependent country within the negotiation process. I sketch and position Bangladesh within the landscape of the process's political blocs and study both its country delegates and civil society representatives to show what leadership looked and sounded like when one's future, as Drs. Nishat and Rahman believed, was already decided.

Bangladesh in the Eye of the Storm

Within the wider climate context, Bangladesh occupies an impossible space, damned if I do, damned if I don't. I won't enumerate here the geographic features and economic statistics that make Bangladesh a country vulnerable to climate change. That information is available from any number of websites, studies, and country profiles. Instead, I present two ethnographic moments to illustrate how often and easily the country was leeched of its complexity because of its perceived climate vulnerability and how its national dialogue about its public perception indicated considerable internal heterogeneity. At a "Rethinking Race in the Anthropocene" conference at the University of Oregon in Eugene in 2015, I distinctly remember the moment when a panel speaker, a Caucasian female climate scientist, started to recount the real-world effects of climate change. I listened with increasing dread as her talk turned to the horrors that lay ahead for vulnerable countries. When she started to speak about Bangladesh, as I knew she would, she began sobbing, saying over and over, "What have they done to deserve this?" The only person from Bangladesh at the conference, I wasn't surprised when heads swiveled to look at me. I sat slumped at my seat, skewered by the force of pity and liberal guilt.

Thankfully, I got approving nods when I pointed out that Bangladesh was not just a victim but a fairly troubled country in its own right, with a problematic government, inadequate infrastructure, no welfare provisions or safety nets for most of its 160 million people, and resource capture by its elites. In the end, I wasn't exactly sure what I had achieved with my bravado other than to assert that Bangladesh wasn't free of politics. Perhaps my insistence only amplified the climate scientist's eulogy for Bangladesh. The country was done for, and its politics only made that inevitability worse.

These politics were in ample view at the "Water, Waves, Weather" conference hosted by BRAC University in Dhaka in 2011 that I attended while carrying out research in Bangladesh. The keynote speaker at the conference was none other than Dr. Nishat, the university's vice chancellor at the time. Not unlike the climate scientist I was to encounter a few years later in Oregon, Dr. Nishat outlined the climate science most relevant for Bangladesh. Correcting the universal perception that Bangladesh's fate was to go under water, he highlighted the fact that soil salinity was a much bigger and more immediate threat to the country's southern coast than inundation. He sketched in stark terms what lay ahead for Bangladesh with every rise of a degree in global temperature. At 4°C degrees of rise, he predicted that there would be no food security. Speaking clearly but with considerable flair—he was, after all, a lifelong educator and a public figure—Nishat said, "We can no longer look to the past to anticipate the future. The future will have to be generated," and he ended with the warning, "We need to adapt or else we are going to suffer." Adaptation discourse was insurgent in Bangladesh at that time.

There was immediate pushback from the audience. Khushi Kabir, the leader of the activist organization Nigera Kori,[1] which sought property rights, legal rights, and other tools for landless people to fight the explosion of shrimp cultivation destroying land and water along Bangladesh's southern coast, asked if it wasn't the case that Nishat's framing of the situation made it entirely an engineering problem to be solved by technical means alone. This exchange, in which Nishat extolled engineering and Kabir expressed her uncertainty toward it, would come to my mind when later Nishat appeared to give up hope even for a technical solution to Bangladesh's problems in the face of the rising waters. Saleem, who was also a speaker at the conference, didn't disagree with Nishat but added what I would learn was a familiar refrain of his, that it was not all bad and that "the world would now come to Bangladesh to learn from it" (Paprocki 2015). While Nishat foresaw a dire future, Saleem anticipated that local and national efforts would join forces with international ones. Kabir remained skeptical of both scenarios.

Representing Bangladesh

I finally approached the Bangladeshi delegation in Bonn to focus on what had brought me to the COPs in the first place: to find out where Bangladesh was positioned within the process. I learned that most of the official delegates hailed from the Ministry of Environment, Forest, and Climate Change, which maintained an exclusive purview over climate-related issues within Bangladesh. In later years, I saw the delegation joined by officials from the Ministry of Disaster Management and Relief, and the Planning Commission. Disaster Management was concerned with how climate was to be brought within the rubric of disaster risk reduction (DRR) principles, which was its bailiwick (see Seddiky, Giggins, and Gajendran 2020); the Planning Commission was interested in the intersections between the climate-related metrics that were being forged at the negotiations and those provided by the UN's Sustainable Development Goals (see Rahman 2021). If there was anything to be learned from this alphabet soup of ministries, principles, and goals, it was the extent to which the administration of Bangladesh hewed to international directives.

By the time I introduced myself to the delegation, I had already attended press conferences thrown by Bangladesh at several COP meetings. They had ranged from a comedy of errors, the minister of environment speaking so unintelligibly about climate change and Bangladesh that people in the audience walked out in the middle of his remarks, to a thorough dissection of COP proceedings by a whole host of Bangladesh delegates who clearly understood the issues down to the finest detail. At COP23 in Bonn, the Bangladeshi delegates were organized, holding daily debriefing meetings at 4 p.m. within the COP premises. Other civil society participants either of Bangladeshi origin and/or whose work involved Bangladesh were allowed to attend, provided they did not interfere with the meeting.

The delegation was largely male, with a few women mixed in, of whom the men were protective and solicitous. The meetings were a fine calibration of attentiveness to the nuances of hierarchy within the bureaucracy in Bangladesh, with those bearing the title of additional secretary, such as of the Ministry of Environment, at the head of the table, flanked by joint secretaries and junior people, including deputy secretaries, senior assistant secretaries, and assistant secretaries, depending on who was in attendance, arranged in order of decreasing importance. When higher-ups joined the meeting, the seating shifted to accommodate them. But when the secretary, usually the head of the ministry, joined them, all efforts at hierarchy fell away, with everyone thronging the minister to attend to his every request.

Outsiders (like me) were either consigned to an outer circle of chairs or to the back of the room. Here too there was some expectation of hierarchy, and people would gesture at me to change my location if I sat down presumptively. There was usually a young-looking official hovering at the door to direct people to their status-appropriate seats as they entered. The meeting commenced with everyone in the delegation being asked to recount what they had done that

day and to relay notable developments and analysis within and outside the negotiations. Subordinates responded to queries; they did not ask questions of
their own, and every response began and ended with "Sir." From the discussions at the debriefings, it became clear that Bangladesh was there as much to
attend side events, initiate contacts, and make bilateral deals as to participate
in the negotiations.[2]

While attention to age, position, and status mattered in the interactions
among Bangladeshi government officials, they were a bit more generically polite when it came to the members of the delegation from civil society, which
included scientists like Drs. Nishat and Rahman; economists like Dr. Mizanur Rehman, who followed finance for the delegation at the COPs; legal experts such as Hafijul Islam, who tracked loss and damage; and members of
government-run development organizations, such as the Palli Karma-Sahayak
Foundation (PKSF), which followed economic opportunities for the government at various side events and country booths. The PKSF was also an accred-
ited entity with the Green Climate Fund and partnered with Bangladesh-based
organizations to apply for funds.[3]

To explain where this ragtag delegation sat within the larger landscape of
country delegations at the climate negotiations, I break from my narrative on
Bangladesh to provide a lay of the land. Since countries were rarely there as
Party delegations alone but also as part of the UN's divisions of countries and
within blocs either representing countries of specific regions or political coalitions of countries, my sketch looks at these various divisions and blocs before returning to Bangladesh.

Divvying Up the World

The UN divvied up the world's nations according to the state of their economies, with "developed," "emerging economies," and "developing" as the salient markers. This division was used in creating lists of countries to go under
the headings of Annex 1, Annex 2, and non-Annex within the UNFCCC. This
division was used for setting carbon emission reduction targets within the Kyoto
Protocol. Developed countries and emerging economies, put in Annex 1, were
given targets. Those developed countries put in Annex 2 were the most advanced economies and were further tasked with providing support of various
kinds to developing countries, with the aim of bringing them to the level of
taking on emission targets. Developing countries, put in the non-Annex category, were not burdened with targets.

Developed countries had subsequently come to resist this early classifica-
tion within the Convention and the Kyoto Protocol because they felt that

developing countries were close to and in some cases had even surpassed the carbon emissions of developed countries and should therefore share the burden of mitigation. This parsing of burden was one of the key reasons for the retraction of Party support from the Kyoto Protocol, leading to its near abandonment. The Paris Agreement, in its turn, discarded the use of Annex and Non-Annex delineations in favor of the appellations "developed" and "developing," with "least developed countries" and "small-island developing states" given special consideration because of the specific challenges climate change posed to them. This reorganization considerably shrank the number of countries needing support.

In addition to categorizing countries by the state of their economies and using this measure to determine their level of participation in the climate regime, the UN also strove for geographical representation, dividing the world into five regions: African Group, Asia-Pacific Group, Eastern European Group, Latin American and the Caribbean Group, and Western European and Others Group (the "others" here include Australia, Canada, Iceland, New Zealand, Norway, Switzerland, and the United States, but not Japan, which was part of the Asia Group). This mode of representation was used when setting the agendas for the meetings of the COP/CMP/CMA. However, it was acknowledged by the Secretariat that the UN's regional groupings and mode of representativeness were not salient for representing Party positions within the negotiations and that there were "several other groupings . . . more important for climate negotiations."[4] These groupings were called political blocs.

Parties Get into Blocs

While the political blocs that constituted the Global North were important players in the process, specifically the European Union, the Umbrella Group (a coalition of non-EU developed countries), and the Environmental Integrity Group, formed in 2000 by six countries that diverged from the positions of the Umbrella Group, I was primarily interested in those from the Global South. The G-77 and China were the most notable among those representing the Global South. When it was created in 1964, the G-77 consisted of seventy-seven developing nations seeking to work together to derive trade benefits and market protections within the global economy. Since that time, the group had grown to include 134 countries, including China. As China did not claim to be a member but articulated its position through the G-77, the group now went by the title of the G-77 and China (see Kasa, Gullberg, and Heggelund 2008; Vihma, Mulugetta, and Karlsoon-Vinkhuyzen 2011; Klöck, Castro, and Blaxekjær 2020).

To shore up its claim that it represented the interests of the Global South, the G-77 and China drew on a long history of political cooperation among nation-states at the periphery of the world system (Prashad 2013). Notable within this history was the 1954 Bandung Conference, at which a number of African and Asian countries met to lay the foundation for political, economic, and cultural cooperation and resistance against incursions of colonialist and neocolonialist powers. The conference served as the precursor to the Non-Aligned Movement, started in 1961 to allow countries to sidestep participation in the Cold War between the United States and USSR (Vihma, Mulugetta, and Karlsoon-Vinkhuyzen 2011), and the creation of the UN Conference on Trade and Development (UNCTAD), designed to secure fair terms for access into the global economy (Toye 2014). UNCTAD would later propose a New International Economic Order to restructure the global economy toward resource redistribution on the grounds of equity; this project was thoroughly routed by internal division and external opposition by developed countries (Toye 2014; Getachew 2019).

The reason for providing this short historical excursus is to explain the importance of the G-77 and China group, which cast itself as the legatee of the movement, protecting the political and economic interests of the Global South at numerous forums, including the UNFCCC. Consequently, its rhetoric was redolent with references to this historic past and calls for solidarity among historically peripheral nation-states (Vihma, Mulugetta, and Karlsoon-Vinkhuyzen 2011). But it was also pragmatic, putting forward only the common stances shared by its member states within the negotiation process. These shared positions within G-77 and China sounded self-evident, such as respect for the original Convention, with its delineation of expectations and responsibilities of developed countries; the call for attention to poverty and sustainable development within developing countries; as much attention to adaptation as to mitigation measures; and finance, technology transfer, and capacity building to better equip developing countries to withstand climate change, to name the major issues. However, they were not as self-evident as they might seem. For instance, the G-77 and China's insistence on the importance of the Convention was not just in recognition of its historic nature, of bringing together developed and developing nations in common cause. It was simultaneously a reminder of the two important principles articulated and enshrined in it, notably "common but differentiated responsibilities" (CBDR) and "equity" (Honkonen 2009), which were sometimes treated as the same. It was with these principles that G-77 and China reminded the industrialized countries of the Global North of their outsize role in creating the problem of climate change in the first place and the importance of taking leadership in mitigating their carbon emissions, while providing flexibility and help to developing countries to combat climate change. These principles correlated with the emphasis on "historical responsibility," "climate debt," and "fair share" within Demand Climate Justice's activist discourse.

The language of sustainable development was also not as self-evident as it might appear. It had evolved to make the actions of poorer nations integrating environmental concerns within their development plans commensurate with those of mitigation by richer nations (Bernstein 2001). And in the case of the Arab countries within the G-77 and China, the emphasis given by the group to adaptation alongside mitigation was not just a reminder of the fact that climate change was already having adverse effects that needed attention but that certain countries would need to adapt both to climate change and to the economic hardships brought about by climate mitigation efforts.

The more I looked into the G-77 and China, the more the sense that their commonsense stances on issues were held in common fell away. Even the appeal to solidarity among the nations within the Global South began to appear disingenuous, given the sharp differences among them in size, the state of their economies, and their current contributions to the global problem of climate change. The one factor that most encouraged smaller countries to form blocs outside of the rubric of G-77 and China, that is, to put into motion schismogenesis, the routes by which divisions occur as a way to resolve internal tensions (Bateson 1958; see also my Introduction), was the emergence of the BASIC group, composed of Brazil, South Africa, India, and China, the four major developing countries that have entered into the emerging category within the UN breakdown of economies. This group was constituted in the 1990s and stepped out of G-77 and China to negotiate directly with the developed countries during COP15 in Copenhagen for voluntary instead of mandatory reduction of emissions (Hochstetler 2012; Blaxekjær 2014; Blaxekjær et al. 2020). This breaking of ranks was terrifying for the smaller countries within G-77 and China, as it suggested that the emerging powers were as indifferent to making meaningful emissions reductions as developed ones. This was to have consequences for new bloc formations after 2009 in which Bangladesh was involved.

In addition to the G-77 and China, there were political blocs of nation-states that sought to focus issues beyond the common positions held by the G-77 and China. There were, for example, the forty low-lying island countries that made up the Small Island Developing States (SIDS). This was one of the UN's original regional groupings, which had evolved into a coalition of nation-states with concerns over their shared plight in the face of climate change. It affiliated with the G-77 and China and, since 1990, with the Alliance of Small

Island States (AOSIS). In fact, AOSIS was created to help consolidate a SIDS-specific voice within the climate negotiations and was unique to the UN-FCCC in so far as it had no other presence or mandate outside of the climate process (see Betzold 2010). Furthermore, it assumed a particular voice, using science to convey the urgency of those facing the imminent destruction of their way of life. Sometimes it crossed over from deploying scientific warnings to speaking prophecy, even sounding as if they were communicating from the other side of the destruction wrought by climate. The ominous quality of this voice was distinctly heard in speeches given by Reverend Tufue Lusama of Tuvalu at various side events on loss and damage at COP23 in Bonn.

The Least developed Countries (LDCs) comprised forty-six countries and was one of the oldest groups within the process, created within the larger UN system, which used population size and gross national product as criteria. It typically operated through G-77 and China in putting forward its interest in adaptation-related issues (Bernardo et al. 2020). However, it started to exercise a little more independence after Copenhagen, meeting with EU, AOSIS, and some Latin American countries to figure out a compromise position with developed countries to avoid a repeat of the failure of COP15, when no decision was adopted at the conclusion of the session (Blaxekjær and Nielson 2015). The African Group of Negotiators (AGN), which also derived from the UN's regional groupings, was created at COP1 in 1995 with fifty-four nation-states in Africa, for the explicit purposes of representing African interests within the climate negotiations and more broadly in the region. Along with LDCs, they made clear their lack of culpability for climate change, the plights of their economies, and their desperate need for finance for adaptation purposes (Roger and Belliethathan 2016).

The two groups, although alike in their conditions and causes of marginality, were quite unlike in their political orientation. The LDC group did not espouse any explicitly political position but manifested its economic dependence on others through its support at one point for the G-77 and China and at other times for the EU. In contrast to the LDC, there was an entrenched, anticolonial sensibility within the AGN. As the young hip spokesman of the AGN, Seyni Nafu of Mali, told me, this anticolonial attitude and vigilance against neocolonialism manifested in the AGN's interest in cooperating among themselves and privileging homegrown experts. While there may be merits to a distinctive African style and mode of negotiation, it is noteworthy that many countries within the AGN also were members of the LDC group. This suggested that the AGN was split or that those Parties (again, "Parties" means "nations") that moved between the AGN and the LDC blocs were effectively moving between two personas.[5]

The Arab League, established in 1945 with a current membership of the twenty-two countries that made up the region, operated the Arab Group within the negotiations. On a mandate from the Arab League, Saudi Arabia had led the negotiations for the past two decades. While most Parties' delegations were composed of officials from ministries of environment or even members of civil society, such as environmental lawyers, the Saudis' delegates were officials from the Ministry of Energy, Industry, and Mineral Resources and individuals seconded by ARAMCO, the Saudi Arabian oil company. Since the early days of the negotiations, the Saudis had been portrayed as "obstructors" within the process, with their motivation clear to everyone (Depledge 2008; Kasa, Gullberg, and Heggelund 2008; Luomi 2011). After all, a world no longer reliant on fossil fuels meant the demise of a large majority of Arab economies. The Arab Group was affiliated with G-77 and China, leading many to claim that Arab nations used G-77 and China to shield themselves from criticism by other developing countries (Barnett 2008). They made the issue of adaptation contingent on including measures to help countries whose economies would suffer as a result of climate-change-mitigation efforts. This controversial issue, referred to by the innocuous-sounding phrase "response measures," rendered adaptation a fraught issue, one marked as important but indefinitely deferred within the negotiations.

The Bolivarian Alliance for the Peoples of Our America (ALBA) and the Independent Association of Latin America and the Caribbean (AILAC) were formed in 2009 and 2012, respectively, to represent Latin American countries within the negotiations (Watts and Depledge 2018; Watts 2020). While both were under the rubric of G-77 and China, they were ideologically different, with ALBA deriving from ALBA-TCP (Bolivarian Alliance for the Peoples of Our America–People's Trade Treaty), the intergovernmental organization working to ground socialist and social-democratic government in an integrated economy in the region. In the negotiations, ALBA had been most prominent in asking developing countries to honor their climate debt. AILAC broke off from ALBA to constitute itself as a group that emphasized building North-South bridges and shared asks with LDCs and AGN (Watts 2020).

The Climate Vulnerable Forum (CVF), which emerged out of the 2007 Malé Declaration on the Human Dimension of Global Climate Change, included forty-eight nations from Africa, Asia, the Caribbean, and the Pacific, transecting the SIDS, LDCs, and AGN blocs. It provided yet another forum for poor and vulnerable countries, specifically at COP15 in Copenhagen. The CVF emphasized that its member states sought to provide both moral *and* economic leadership in greening their economies. However, to do so, they needed to be assured of public finance to the tune of 1.5 percent of the gross domestic product of developed countries, above and beyond the current grants of 0.7 percent of GDPs given in global aid by the Organization for Economic Cooperation and Development, composed entirely of developed countries. While some commentators felt that the proliferation of blocs such as the CVF was important in putting moral pressure on developed countries to provide financing to help developing countries fight climate change, others pointed out that the shift in rhetoric from moral to economic leadership was telling in that it suggested that the smaller, poorer, and more vulnerable nation-states within the G-77 and China were seeking to shame the larger developing countries in their bloc into doing more for the global effort (Blaxekjær and Nielson 2015). They said that the CVF was the product of developed countries "peeling off" nation-states from the G-77 and China bloc to undermine what fragile solidarity and consensus remained within the Global South.[6] The total number of states within the LDCs, SID, and/or CVF was considerably fewer than the total number of developing countries, which made the economic burden of helping them much lighter than helping developing countries as a whole. The CVF seemed to have a fluctuating presence within the negotiations, being publicly present during COP21 in Paris but less so since.

Just from this discussion of the major groupings I hope I have been able to indicate how difficult it was to imagine that G-77 and China had captured the mantle of leadership of the Global South and to what extent it was perceived to have failed vulnerable countries (SIDS, AGN, LDC) for having been instrumentalized by others (Arab Group, BASIC). But before we think all is lost, let us consider one final group, one that had come into existence in recent years, whose approach might not appeal to the liberal in us asking for immediate and decisive climate action but that will allow us to appreciate the complex ways a Party multiplied its positions and impacts through its affiliations with political blocs.

The Like Minded-Group of Developing Countries (LMDC) announced its emergence in 2012 (Blaxekjær et al. 2020; Halkyer 2021). Among its members was an unusual assortment of countries that would not consider themselves allies, such as Saudi Arabia, Iran, Bangladesh, Pakistan, India, and China. At any rate, its membership was variable, with countries coming and going, announcing their alliance with this group by speaking on its behalf in negotiations but then speaking from a different position in other years (Bangladesh being one of those). LMDC insisted that it was anchored by the G-77 and China, and it demonstrated this by backing the one unanimous position within the G-77 and China, that is, the espousal of love and commitment to the Convention, specifically to its principle of CBDR, which meant bringing equity into consideration when distributing responsibilities for mitigating climate change.

This principle was enshrined in the Framework Convention and traveled from it to other environmental agreements. In practice, it meant maintaining the distinction between Annex and Non-Annex countries as laid out by the Convention, with Annex including developed countries and emerging economies and made responsible for taking the lead in climate action and Non-Annex meaning developing countries, including least developed countries and small-island developing states, for whom climate action was voluntary and optional. This understanding of CBDR was upheld in the Kyoto Protocol, but, as I will show later, it did not hold for the Paris Agreement, at least not quite in the same way. The new phrase within the PA was "common but differentiated responsibilities in light of national circumstances" (CBDR-NC). This addition of "in light of national circumstances" could mean that one might not have the national circumstances to take on burdensome climate action and that these constraints would be kept in mind when evaluating what one had done for climate action. But it could also mean that those with national circumstances that had improved over time could be upgraded to bear more climate-related responsibilities.

The countries within LMDC did not put forward any other common positions than that of CBDR and related mandates. They refrained from utilizing this platform for national interests. Rather they repeated their commitment to CBDR to the point that the repetition started to grate. One couldn't help but call them blockers, shielding intransigent tendencies and rogue nations within the climate negotiations.[7] But in my reading, the LMDC maintained the solidarity of the Global South, keeping alive the historical promise of G-77 and China to protect the weak of the world while the latter group was compromised and split within itself. It suggested how schismogenesis is not entropic, that is, leading to greater and greater disorder and chaos (see Roberts 2011), but that it can also produce reconsolidation into new formations (Bateson 1958; Prigogine and Stengers 2018). One hears this hope in the words of Gary Theseira, the deputy undersecretary of the Environment Management and Climate Change Division of the Ministry of Natural Resources in Malaysia, who spoke on behalf of LMDC at the closing plenary of the COP in Katowice in 2018. At that time, the final decision text that was to make up the Paris Rulebook had been approved, and Parties were giving speeches congratulating one another on having come up with guidelines on how countries were going to implement the agreement. It is worth quoting him to give a flavor of LMDC's rhetorical heft:

> It is crucial to recall why developed countries are developed and we are developing. Historically, the largest share of the carbon/emission

space has been used by developed countries in Western Europe and North America as they industrialized fueled by the use of fossil fuel—and their economies prospered. This was the genesis of climate change. Developing countries are also entitled to develop and grow sustainably. For this, the remaining carbon space must be shared equitably. This is the premise of the Convention and its progeny— the PA. We owe this to the poor and vulnerable who are paying— sometimes with their lives in our parts of the world. This is why equity assumes a major and central role in meeting the climate change challenges. That is why equity must be operationalized and meaningfully . . . (Raman 2018a, 3).

Leading as Bangladesh

In keeping with the UN's custom of upholding national sovereignty, the climate negotiations sought to give countries the sense that their participation was equally weighted. Thus, within each official negotiation room there was always a table to the side with placards for every country in the process. As soon as a negotiator entered a room, they went and found their country placard, placing it on the table before seating themselves behind it. When they wanted to speak, they flipped their country placard up. When Parties spoke, they would say their country name and that they were speaking on behalf of a bloc with which they were affiliated. In fact, depending on the circumstances, they most often spoke as both a Party and a bloc, rarely as a Party alone.

The presiding officers kept reminding those gathered during negotiations that it was a "Party-driven process," hinting that they would rather hear an individual Party's position than one mediated by a bloc. Sometimes they even chastised Parties by indicating that their positions appeared incoherent, dispersed across too many blocs. When this happened, the officers expressed anxiety not only that a Party might have lost its way, but that it might be ventriloquizing the voices of those absent from negotiations who, nonetheless, had a long reach. While developed-country Parties had reams of people in their delegations to oversee negotiations on every issue, smaller, poorer countries did not have such numbers in their reserves (Chan 2021b). They instead often relied upon the proffered help and expertise of others, making it uncertain in whose voice they spoke.

Bangladesh had standing within the negotiations, both in so far as it had official recognition like the rest and because it showed clear leadership, just as Saleem had told me. Prime Minister Sheikh Hasina was given an award as "a champion of the earth" by an UN entity in 2015 (*UN News* 2015). Bangladesh was also chair of the Climate Vulnerable Forum bloc from 2011 to 2013 and resumed chairmanship of the forum in 2020. Yet there was a difference between the early years of Bangladesh's participation to these more recent ones. As part of the G-77 and China, it had once advocated for CBDR, even being part of the fiery LMDC bloc for a while. But as LDCs became more confident of their own position within the negotiations apart from G-77 and China, Bangladesh came to associate more with them. It adopted the truculent voice I associate with the LDCs, the one that spoke of the need for adaptation, climate finance, technology transfer, and capacity building, what I came to think of as the "four asks," versus CBDR. But this voice changed once more in the 2010s, when Bangladesh seized on loss and damage, an issue that had traditionally been associated with SIDS-AOSIS to force developed countries to mitigate, which had a different kind of fierceness than the demand for CBDR. At the same time, Bangladesh maintained its foothold within the LDCs, transforming its truculence into moral authority tied to specific economic proposals through its leadership of CVF.

While its leadership was not disputed (nor were such fluctuations of bloc affiliations and contradictory positions out of the ordinary within the process), official and nonofficial Bangladeshis spoke in a polyvalent way, maintaining several distinct and seemingly contradictory voices within the COPs (Bakhtin 2013). Sometimes it was not clear that Bangladesh's voice was necessarily its own. This is not to suggest that it was someone's puppet, but the aspect of its overidentification with developed countries made the speech of its delegates sound somewhat ideological. The ordinary language philosopher Stanley Cavell (1994) has explored how threatening it is for one's sense of self to be told that if one keeps breaking one's promise, one will no longer be believed. Yet within this process, countries of the Global North had repeatedly broken their promises, and there was no sense of such a threat to one's selfhood or its attendant feeling of shame. Rather, shame was imposed on those forced to pawn their voices, as then they were continually confronted with the question of in whose voice they were speaking.

Once the Bangladesh delegations' daily debriefings ended in Bonn at COP23 and the delegation dispersed, all deferential behavior was shed, and the delegates became part of the process. I was invited to shadow Ziaul Haque and Mirza Shawkat Ali, the director and deputy director, respectively, of the Department of Environment within the Ministry of Environment, Forest, and Climate Change. The two men must have been in their early forties. I called them "bhai," or brother, to show my respect for them, and they called me "apa," or big sister, reflecting my age relative to them. The nickname also expressed their increasing affection for me, whereas their boss, Dr. Nurul Quadir, never referred to me as anything other than "madam," which served to keep me at arm's length.

I first spoke with Dr. Quadir, from the Ministry of Environment, Forest, and Climate Change, about Bangladesh's priorities. It was my introduction to the country's strongly neoliberal orientation. By this I do not only mean a positive orientation to the market economy but also a sense that one alone is responsible for one's state and eventual fate, which foreclosed analysis of the historical and structural causes of underdevelopment (see Thorsen 2010). Dr. Quadir felt that the convention was a genuine opportunity for a country like Bangladesh because "UNFCCC was a process, it wasn't a top-down, one-size-fits-all development model." Through this process, relationships stood to be created. While Bangladesh was clearly on the "development pathway," it needed to move away from pursuing development funds and instead seek investments. Dr. Quadir thought that climate was an opportunity to address the investment gap, that "climate was another business opportunity."

When I heard Dr. Quadir talking like this, I couldn't help but recall Khushi Kabir's words to this effect back in Dhaka, that Bangladesh was addicted to development funds and that climate was just another way to tap the developed world. However, Dr. Quadir was being more nuanced than this. When he said that climate was a business opportunity, he meant that it was an opportunity for Bangladesh to do some "soul searching," to take another look at "the products it had on offer" and "the expertise that it counted on," to ask itself, "What is my capacity?" "What do I need to achieve?" and "What relationships will help me achieve these goals?" Further on, he said, "They [meaning the developed world] need us to do something. They are dependent on us." While this reversed the usual relationship of dependency, it was a different reversal than that produced by the concept of climate debt. In Dr. Quadir's picture, the developed world *depended* on the developing world to help out with mitigation, whereas within climate debt the developed world *owed* mitigation to the developing world.

Thus far I had been thinking that sustainable development and poverty alleviation were the means by which developing countries could contribute to fighting climate change. Adaptation was later added to this list as their equivalent of mitigation by developed countries. Through Dr. Quadir's words, I was coming to understand that the Paris Agreement made mitigation into a new opportunity for Bangladesh. It was the opportunity to join the big guys, to think in their terms, and to develop relationships to enable climate action in their terms—and by which to benefit in ways yet to be explored.

Following the Lead of Negotiators

While I wandered around in Paris in 2015 trying to wrap my head around the immense enterprise called the COP and followed activists and various civil society organizations in Marrakech in 2016, thereafter I listened in on as many negotiations relating to mitigation as I was able to access. In each subsequent session of the COP, in Bonn, Katowice, and Madrid, I shadowed Zia bhai. While my access to actual negotiations was spotty, one could only enter either with one of the few tickets provided to one's constituency (mine was RINGO), or if the negotiations were open to all observers, I was able to be present with Zia bhai at a number of meetings that were out of the general stream of things, having to do with the contingencies of the moment, the fast devolution of issues, and the need for on-the-spot responses.

Zia bhai was following the Ad Hoc Working Group on the Paris Agreement's (APA) Agenda Item 3, which had to do with operationalizing mitigation within the Rulebook. This item was where the action was in Bonn and Katowice. As an indication of how resourceful his superiors found him, Zia bhai was also following issues under consideration by subsidiary bodies on behalf of the LDC group. In addition, he was attending the negotiations on the Consultative Group of Experts, of which he was a part and whose mandate was to help developing countries with their reports to the UNFCCC. Its tenure was soon up, and he, along with others within it, was seeking to extend its work plan for another eight years but was facing resistance from the United States and Saudi Arabia, who sought to scale back the UNFCCC. I sat around while Zia bhai strategized with a few other members of the CBE and watched in admiration as he diplomatically battled the Saudi delegate seeking to nix its existence. What was impressive was that Bangladesh felt empowered by its standing within the LDC group and experience with international work to be able to push back against the United States and Saudi Arabia within this space, albeit in a limited capacity.

Zia bhai crackled with energy as he walked into a meeting room, gripping his leather bag in his hand, finding his country placard, and positioning himself next to Selamawit Desta Wabut, the young legal expert with the Ethiopian delegation. The two together represented the LDC bloc on mitigation-related issues. He always listening intently, making statements as needed on behalf of the LDCs, before rushing off to another meeting. As we walked and talked, Zia bhai would explain the different parts of the process to me, or tell me about his personal background, or give me his read on the ever-shifting landscape of politics, as much interpersonal and bilateral as impersonal and multilateral.

When Zia bhai tired of me following him around, peppering him with questions, he offered to have me follow Mirza Shawkat Ali. To my surprise, I learned that Shawkat bhai, who I had thought was mostly engaged with maintaining protocol at the Bangladeshi debrief sessions, since he was always positioned at the door ensuring everyone sat where they should, was the second-most-influential delegate of the lot, charged with following adaptation on behalf of Bangladesh, although not on behalf of the larger LDC bloc.

Shawkat bhai carried his tall thin frame gingerly and seemed affected, even appalled, by the press of people and smells around him. He explained that he had developed a mysterious ailment over the past few years and as a result had lost a lot of weight and felt exhausted. Whereas once he and Zia had been negotiators extraordinaire (he showed me photos of himself from the UNFCCC collection in better physical form, during intense negotiations), Zia now did most of the negotiating. This was likely because activity on the issue of adaptation was almost at a standstill after its acceptance within the pillars of climate action. Zia bhai was protective of Shawkat bhai, ensuring that he got regular meals (purchasing food was always a struggle at the COPs given the volume of people attending) and accompanying him back to their lodging in the late afternoon to rest before returning to continue negotiating late into the night. Shawkat bhai hoped to bounce back soon, speculating that by then adaptation would once again be of central importance.

While it was Zia bhai who explained the minutiae of the process to me, it was Shawkat bhai who gave me insight into what Bangladesh needed to achieve in this process. He agreed with Saleem that while Bangladesh's fate might be tied up at home, it achieved a lot through its leadership of Climate Vulnerable Forum. Shawkat bhai told me with pride that Bangladesh's days of begging with a bowl were over. It was graduating from LDC status to that of an upper-middle-income country, according to criteria set out by the World Bank, thanks to its booming economy. And, echoing the words of the forum, he said that Bangladesh was there to provide both moral and economic leadership among developing countries. The country would not wait to be told to mitigate and adapt. It would leap to do so with whatever resources it could muster. In so doing, it would demonstrate leadership and, simultaneously, put to shame larger developing countries that were whining that developed countries should take the lead.

Here he was referring to the insistence by G-77 and China and the LMDC on CBDR and equity. And he dismissed the four asks, adaptation, finance, capacity building, and technology transfer, considered sacrosanct by blocs such as the AGN and Arab Group, as pie-in-the-sky. Why should developed countries give up what they had developed on their own to others? Why didn't countries feel shame demanding that things be given to them for free? Why couldn't they develop their own capacity, technology, and finance? He pointed out that Bangladesh already had more funds coming into the country than could be gotten from any of the funds under the financial mechanisms within the UNFCCC. Bangladesh was not there to secure funding from these entities; it was there to build bilateral relations to generate new economic opportunities. I had noticed that only a few members of any given delegation were focused on the actual negotiations, while the vast majority was in exploratory conversations taking place across every part of the COP.

Polyphony at the Press Conference

Growing up in Bangladesh, I often cringed hearing the various prime ministers and ministers give passionate speeches at world forums asking for help. Hearing Shawkat bhai speak, I was thrilled at the fiery speech coming out of such an enfeebled body, at the image of Bangladesh ceasing to present itself as needy, dependent, and vulnerable. But the voice of plaint was never too far behind. I heard it as early as 2015 at a press conference by Bangladesh at COP21, when Zia bhai and Shawkat bhai described Bangladesh at being at the top of its game within the negotiations. At the press conference, spokespeople for Bangladesh said in no uncertain terms that while "Bangladesh used to be the land of floods, now it was the universal victim of climate change." Furthermore, adaptation was necessary for Bangladeshis to survive, and for that they needed funds in the form of grant money, not loans at concession rates.

As the sociologist Kasia Paprocki (2015, 2018) has remarked, Bangladesh, one of the most market-friendly countries in the world, had made climate change adaptation into an economic opportunity. And as the anthropologist Jason Cons (2020) has pointed out, Bangladesh has managed to maintain the contradictory image of being a country at the brink of ruin from climate change and, simultaneously, a country flush with economic opportunity. I was seeing these national trends play out at the COPs: The Bangladeshi spokesperson at the press conference declared Bangladesh to be a victim, Dr. Quadir spoke of climate as a business opportunity, and Shawkat bhai spoke with pride of Bangladesh being upgraded to upper-middle-income country status.

These contradictions weren't limited to the official delegation. They were there among the civil society activists as well. While shopping in Madrid for souvenirs for every member of his office back home in Cox's Bazaar, as well as for some makeup for his daughter, the youthful Mizanur Rehman Bijoy from the environmental group Coastal Development Partnership said to me, "Apa, you know Bangladesh. You know we need all the help we can get. That is why I come here, to draw these people's attention to our efforts to make our people more resilient, so they can fight the rising waters."

Meanwhile, at one of the press conferences, Tanjir Hossain, of Action Aid, one of the most forward-thinking NGOs in Bangladesh, which had incorporated climate into every aspect of their projects, shook his head at the plaintive note struck by the spokesperson about Bangladesh's need for grant monies. He turned to me and said urgently, "That's not what we need; we need energy." In a fierce tone he told me that "poverty was pollution" and that "growth was inevitable," and he proceeded to give me the distressing news about Japan's plans to build coal-fired plants in coastal Bangladesh, which started in 2013 and is still ongoing as we speak, entailing enormous financial and environmental costs (Goso 2020), and Russia's plan to build two nuclear power plants in northern Bangladesh to shore up its energy production, which was also supposed to have started in 2013 (WNN 2020). When I asked him how that was helping climate change, he said, "We first need a carbon footprint to be able to do mitigation."

Some version of this position showed up again at an event by Bread for the World titled "Towards a Just Energy Transition" at COP23 in Bonn. The Christian organization was involved in bringing civil society members from developing and least developed countries to the COPs. At this event, Jahangir Hasan Masum, also of the Coastal Development Partnership, gave one of the fiercest speeches I had heard from a Bangladeshi within this space. He said that the need was not just for energy transition, which assumed everyone was energy rich and it was just a matter of making the right choices to ensure that one's energy was ethically sourced. Large segments of the population in Bangladesh remained energy impoverished. To thundering applause, Masum barked that people needed energy and energy sovereignty before just transition. I wondered if the well-intentioned people in the audience suspected that he might be referring to Japan's and Russia's plans for Bangladesh.

This combination of climate catastrophe and business opportunity and the mixture of lament and proud appeal to independence among the Bangladeshis at the COPs were not so much contradictory as polyvalent. It pointed to the fact that no one individual held both views but rather that both points of view existed within the same domain, aware of each other and not necessarily troubled by the other. This polyvalence allowed Bangladesh to navigate the COPs without needing to construct and police a single image for itself.

Reconsidering Bangladesh's Leadership

In addition to leading as a country, the Bangladeshi delegates also led within the process itself. I learned from Shawkat bhai that Bangladesh almost always took the lead among the LDCs in terms of meeting reporting asks. For example, he pointed to when the UNFCCC started requiring National Adaptation Program of Action reports after adaptation was installed as a second pillar of climate action within the process. Bangladesh designed and completed its well before other countries had even figured out the template for the report. It had also completed a countrywide Climate Change Strategy and Action Plan on its own. Both reports were regularly updated. Furthermore, Bangladesh had been providing National Communications to the UNFCCC and had also submitted an Intended Nationally Determined Contribution ahead of the Paris Agreement, both optional for LDCs. Shawkat bhai recalled that he would show up at the negotiations with the reports fresh off the press but run out of copies within a day or two, because fellow delegates took them to help guide their own countries' reporting. I have already mentioned how Zia bhai took the lead within negotiations on behalf of both Bangladesh and the LDC group.

Shawkat bhai further recounted how in the early days of loss and damage, Bangladesh pushed not for liability or compensation, the two issues that made loss and damage anathema for developed countries, but rather for a settlement policy or agency to coordinate the emerging issue of climate-displaced refugees. I had often heard Saleem prophesize that climate refugees, increasingly inevitable, were going to embody the failure of this process. Although the idea for such an intergovernmental body to oversee feared migration and resettlement was put aside, Dr. Quadir was involved in the hard-won Warsaw Mechanism on Loss and Damage, the clearinghouse of information on best practices for how to deal with loss and damage. This mechanism had evolved out of dissatisfaction to the current response to the demand for more attention to the fact that climate change was already wreaking havoc on people's lives. This was another instance of how Bangladesh provided leadership within the process.

While we might say that the Bangladeshis were largely there to be there, nonetheless, the process was an opportunity for the country to practice at being the first, taking initiative, being a good sport, and playing along. It was a position that gave it pride and self-respect. However, it also made it seem like a pawn in the hands of those who gave them development funds, such as the EU. Regardless, the Bangladeshi performance of self-possession and leadership while being dependent was an intriguing one. It will be further parsed out in upcoming chapters to draw out not just the soft power and moral authority practiced by countries such as Bangladesh but also to show it had to ground and affirm itself in the process of venturing into the new territory of loss and damage, which was endorsed neither by its development partners nor by the G-77 and China.

3

Who Wants to Be a Negotiator?

On the flight to Glasgow to attend COP26 in November 2021, I overheard two men speaking about the COP. The plane was likely filled with participants going to attend the session, so I was not surprised to hear reference of it. What surprised me was the men's affection for the event. Said one to the other, "I love this process; attending the COPs is the highlight of my year." I leaned in to learn more, but they dropped their voices when they likely caught sight of me.

Later, I would make it a point of introducing myself to the man on the plane I'd attempted to eavesdrop on. As he asked that he not be named, I will merely identify him as a member of a delegation of a small country in the Global North that was part of the EU bloc, referring to him as the "delegate in Glasgow." The delegate described himself as a latecomer to the process, having trained in the biomedical sciences and environmental toxicology, entering the diplomatic sphere only after pivoting to work on greening the economy, taking a job at his country's Ministry of Environment, Climate Change, and Planning. He described Michael Zammit Cutajar, the UN official of Maltese origin who shepherded the intergovernmental deliberations that led to the drafting of the UNFCCC, as a hero of his whose career he followed avidly and whose exemplary diplomacy he strove to emulate. "Cutajar is interesting because he asked to work on climate at a time when no one else was thinking about it. It was like a hobby of his, but without him taking up this hobby we wouldn't have made what little advances that we have in terms of having a global process that brings everyone together to fight climate change as a threat to all humanity."

But had the process delivered, I asked. He confessed that climate change was only now becoming an issue for most countries, and that, too, as something

to attend to only occasionally, and that for the most part it was seen as someone else's problem. He said, "It's not in our backyard," "the climate is an
emergency but not emergency enough," and, later, "adding up events doesn't
create urgency." But progress in the process was more specifically hobbled by
the fact that policy could only be made by consensus. He reminded me that
without adopting rules of procedure, whose adoption had been "blocked" by
Saudi Arabia at the start of the process in 1991 because of its objection to majority voting, the process could only proceed if everyone agreed on everything
(Kemp 2016). Achieving consensus on every issue held back the process, from
his perspective. However, as I later reflected, this one requirement, arguably a
major block to swift climate change, was also likely leveraged by countries in
the Global South to stay relevant to the process. "At any rate," he continued,
"in a world in which every country's sovereignty is dearer to it than anything
else, the COP is the best thing we have. . . . It is a strange process. It works in
mysterious ways. It's not clear how the process is progressing, but it is best to
let it continue as it is, or else we will likely be doing nothing." Maybe these
words depicted the process as more mysterious than it was, but they captured
a general sense that it succeeded by keeping the countries of the world on topic
and in conversation.

I asked a question of the delegate in Glasgow that had long bothered me:
What did it mean to be a negotiator if one's country's positions, and even its
red lines, on various topics were already publicly known? How could one be
taken seriously if others already knew the extent to which one was allowed to
go? He replied, "No delegate comes with just a straight 'no' on their pad. On
my side I have my needs, which are critical, and for which I will work tirelessly.
I have my wants, which are those things that would be nice and good to have.
And then I have my likes, which are not important at all. The other side has
these as well. I just have to figure those out and figure out if a deal can be cut.
At the end of the day, that is negotiation." As he said frankly, it was the "thrill
of the deal" that endeared the process to him.

I learned a lot from this delegate about how at least some in the Global
North saw the process, and I appreciated the candidness with which he spoke.
However, our conversation reaffirmed for me why I had made the right choice
in focusing my study on the Global South. I could see that the COP was an
opportunity for thrilling deal making. I could see how representatives of the
South could be equally caught up in that thrill. Recall Dr. Quadir's assertion
that Bangladesh sought to be involved in the myriad economic opportunities
opened by climate change mitigation, to also be in on the game. At the same
time, I could see how leveraging climate change as an economic opportunity
ran the risk of diminishing the depiction of the suffering of humans and non-

humans spoken about in the sessions, of words losing their connection to reality (Das 2006). Even if the Global South was heterogeneous, disunited, even sabotaging of one another, or precisely because it was, this was the way we were forced to contend with climate change: This messy politics was climate change in another form, prefigured by it, experiencing it, and now anticipating its worst effects.

In this chapter, I speak with negotiators within country delegations to explore their pathways to the negotiations and how they viewed negotiation as their vocation. I observe training sessions of negotiators of the Global South to identify what is or is not achievable within negotiated texts and to gauge the power of deal making and deferrals. I track the many-striped experts who provide guidance to delegations from the Global South to show how they attempted to influence the negotiation process.

Negotiation as Vocation

The delegate in Glasgow went to the COPs because it was his profession, but the sessions also gave him much enjoyment. It was as if negotiations were in his blood and he lived for them. This feeling made negotiations a political vocation for him in the Weberian sense of the word, in which one's work at the COPs was felt to serve a higher cause (Weber 2004). He had started as a scientist, who were expected to eschew politics and were chastised if seen to be partisan (Callison 2014). We know that between the 1990s and the 2000s climate change went from being a *problem* for scientists requiring the integration of knowledge across different lines of research to being a *cause* for politicians (Edwards 2010). How did people with scientific backgrounds, specifically those from the Global South, shift to becoming negotiators?

A response may be sought in how the UNFCCC established itself within the architecture of individuals' lives. In the early years of organizing, from the 1970s to the 1980s, the UN relied on institutions such as the World Meteorological Organization and later the Intergovernmental Panel on Climate Change to amass credible scientific data to ground claims of climate change and its links to human activity. When the climate process was first launched by the Rio Summit of 1992, it relied heavily on science to communicate the urgency of climate change to nation-states to organize them to collectively combat it (Lanchbery and Victor 1995). By the 2000s, climate change was accepted as already here in the world, and the COPs seemed powered by its own momentum and showed less and less reliance upon the IPCC to guide its policy-making initiatives (Livingston and Rummukainen 2020). And by means of participation in these intergovernmental exercises, an entire generation of country delegates from the Global South who came from a science background came to be entrained within the negotiation process.

I provide examples of three individuals I interviewed in 2018, each the head of a political bloc within G-77 and China, to draw out how the negotiations were framed in their lives. Gebru Jember Endalew, a Party delegate of Ethiopia and the chair of the Least Developed Countries (LDC) group between 2017 and 2018, recounts that his advanced training was in meteorology and air quality. He came to the issue of climate change via a circuitous route, beginning as a meteorologist in Ethiopia's National Meteorological Agency, then in the Department of Development, from there to the Department of Weather and Early Warning, and finally to the Climate Change and Air Pollution Studies Team. He represented Ethiopia within the IPCC, where he was one of the authors of the Working Group I Physical Sciences Basis of Climate Change Report in 2013. He readily admits, "I was much more at ease with approaching climate change as a scientific problem. I had to learn to think about climate change as a political problem before I could be effective in representing Ethiopia within the negotiations to which I had come in 2008." Negotiator training offered by an international organization working to ensure that delegates from the Global South were not run over by their northern, wealthier counterparts helped him find his voice and eventually lead the Least Developed Countries in the negotiations.

Amjad Abdulla was a delegate of Maldives, chair of the Alliance of Small Island States (AOSIS) group between 2015 to 2018, and at one point the vice chair and chair of one of the two permanent subsidiary bodies under the UN-FCCC that gave scientific and technological support to the negotiators. He entered the process in the early 2000s as a young man from the Maldives Department of Climate Change and Energy Department, which was housed incongruously within the Ministry of Housing and Environment. "Maldives has always engaged with [the issue of] climate change, so it was like tradition to come. I came with no clue as to what to expect." Armed with a higher degree in environmental science, policy, and planning and an unshakable sense that this process somehow mattered, he came back year after year, with no clear direction in mind but to listen until he found his bearings. He now felt that the process was a "family." Even so, it was all too vulnerable to geopolitics, but "you don't want geopolitics to enter the process. You want to secure it. Good politics is to keep the process technical. Keeping it technical will make it sustainable, whereas making it political will give it a short time frame."

Seyni Nafo, introduced earlier in my description of blocs, was high representative of the president of the Republic of Mali for climate, which is to say that he didn't come via a ministry but was sent by the highest office of the country; he also served as a spokesperson for the African Group of Negotia-tors (AGN). He entered the process "almost by accident." His parents were international civil servants. He was raised in Saudi Arabia and from early on studied finance. When he returned to Mali in 2008, "carbon was very hot then, with developing countries thinking that they could raise a lot of money that way." Seyni started an investment fund dealing in carbon. A presentation to Mali's Ministry of Environment led to his being put in the delegation to do side events at the COPs. "I had no background in diplomacy, which was prob-ably good, as it would have made me too formal," he says. He quickly became integrated into the process, organizing among the African nations so that they could present a common platform at the meetings and undertaking regional initiatives, such as the African Renewable Energy Initiative, an Africa-owned and Africa-led effort to make the region energy independent. "For me," he said, "this process is all about building solidarity."

Endalew had to change his perception of climate change from being a sci-entific problem to thinking of it as a political one to learn to lead in the nego-tiations. We also see that he received training in negotiation to help him make this transition; I will provide an ethnographic account of one such training session later in this chapter. In the case of Abdulla, he attempted to protect the negotiations from geopolitics by framing it as a purely technical process.[1] He wasn't trying to deny that climate change was a political problem that needed recognition by the international community of nation-states and their political will to produce solutions. But it was "good politics" to keep the focus of the negotiations on procedure, regardless of nation-states' internal politics and interrelations. And in the case of Nafo, who was an entire generation younger than both Endalew and Abdulla, we see that he did not even need science to find a pathway into the process; his background in finance had pro-vided it. Although corporate finance was his métier, he leveraged it to become AGN's voice for climate finance for the developing world. Furthermore, he also used it as an opportunity to build regional solidarity and energy sover-eignty. For better or for worse, for climate change to become a part of the ar-chitecture of these technocrats' lives, it didn't just have to depart the realm of science; it had to enter the realm of profession and vocation and become the occasion for new relations and possibilities for solidarities.

Bloc affiliation and the specific identities of the blocs also found expres-sion in individual lives and modes of self-representation. Endalew was affili-ated with LDCs and as such came late to negotiator training and experience in his life, mirroring LDCs' late emergence as an active bloc within the process aided by international experts (Blaxekjær and Nielson 2015). Abdulla, who was affiliated with AOSIS, highlighted its science-driven focus when speaking of the negotiations as a technical process (Betzold 2010), while Nafo expressed the AGN's interest in climate finance and push to emphasize solidarity both to signal that African nations were now united after many years of disunity in their ranks and to elicit solidarity from other developing nations (Roger and Belliethathan 2016; Chan 2021).

Training within Blocs

Before each COP and midyear intersessions, G-77 and China, SIDS-AOSIS, AGN, and LDCs were allotted days and space by the UNFCCC Secretariat to enable Party delegates to hold preparatory meetings. Other country delegations and political groups held similar meetings, but they made their own arrangements. These provisions for G-77 and China, SIDS, AGN, and LDCs were to help redress the unequal distribution of wealth in their midst and ensure global participation in the negotiation process. The Secretariat even paid for up to two delegates from certain countries to ensure their participation, although of course they would remain at a structural disadvantage, given that country delegations could number in the hundreds.[2]

Each bloc was given two days to meet, with the subgroups (SIDS-AOSIS, AGN, LDCs) meeting first, before they all met under the banner of G-77 and China. Each subgroup designated a lead negotiator for each of the topics (specific themes or agenda items) to be negotiated within the upcoming meetings, such as mitigation, adaptation, loss and damage, finance, gender, transparency, etc. The lead negotiators within the subgroups represented their blocs within G-77 and China, which selected its own lead negotiators to represent the bloc within the negotiations. However, sometimes if there was divergence between the position of a bloc and G-77 and China, then the bloc represented itself. Zia bhai, for instance, represented the LDCs on mitigation, as LDCs did not hold a common position on mitigation with G-77 and China. These preparatory meetings were very important for coordinating positions and negotiation strategy across the blocs. Each also held a coordination meeting every day of the official sessions. Such daily coordination meetings were announced on the CCTV monitors even though they were closed to outsiders.

Parallel unofficial events run alongside these pre-COP preparatory meetings. I attended the DCJ-organized two-day assembly before COP23 in Bonn in 2017 and the People's Climate Summit that ran concurrently with COP23. As I started to research the blocs making up the Global South, Saleem told me about the LDC negotiator training run by the International Institute for Environment and Development (IIED), the UK-based research organization of which he used to be a part and whose climate change cell he helped found.

In fact, he told me he had helped broker the partnership between the LDCs and IIED by which LDCs came to receive research support from IIED on negotiation issues and process.[3]

So it was through Saleem's introduction that I found myself in the day-long training in early November 2017 in Bonn undertaken by the European Capacity Building Initiative (ecbi), a joint project with IIED to "build and sustain capacity among developing country negotiators" and "foster trust between both developed and developing country negotiators."[4] It was at this meeting that something about what was meant by a developing-country Party and participation as such within political blocs and the wider negotiation process became clarified.

The 2017 pre-COP training workshop was at a hotel close to the official site of the COP. Thirty-nine Party delegates from developing countries attended the workshop, staying almost the entire day, enjoying the generous tea breaks and lunch and ending with a jolly dinner. A few kept slipping out of the conference room for urgent phone conversations in the hallway, but most others stayed put, indicating that they were here to get as much out of this learning experience as possible—or perhaps they were low enough in the hierarchy of their Party delegations that they had nowhere urgent to be. Quite a number were in their mid-twenties and female.

The COP was starting in two days. The news that the United States issued an official request to withdraw from the Paris Agreement was very much on people's minds. That first day of the negotiators' workshop there was also going to be a citywide climate march led by local and international activists urging prompt action. I didn't get any indication that anyone in the workshop was even aware of the march. The workshop was dominated by government officials, whom I found to be generally disconnected from activism, even by those of their own country. The opening remarks were given by figures important within the COP, some behind the scenes and others in more public ways (ecbi 2017).

Dr. Benito Mueller, the head of the Oxford Policy Institute, which brought together senior negotiators from developing and developed countries in conversations at Oxford University to produce trust among them, kicked off the workshop by reflecting on the historical emergence of ecbi and the resonance between that earlier moment and the present. He recollected that ecbi emerged in 2001, after the United States withdrew from the Kyoto Protocol. It aimed to facilitate Europe taking on leadership of the climate process with support from developing countries. And now again ecbi was meeting, just as the United States was withdrawing from the PA and with a renewed call for Europe's leadership in helping developing countries. Mueller spoke with the easy assumption that to grow trust and efficacy within this process, developed and developing countries needed to know and sympathize with one another, and what better way to do that than by face-to-face contact and individual relations. Upon reflection, I found Mueller's position very optimistic, as it seemed a denial of fundamental power asymmetries. I fretted whether this was another way developed countries could influence developing countries.

Ambassador Luke Daunivalu, deputy permanent representative of Fiji to the UN, followed Dr. Mueller in giving the opening address. Fiji was the first SIDS country to preside over COP23, which was being cohosted by Fiji and the city of Bonn. He put forward the Talanoa Dialogue as Fiji's main contribution to the COP, by which Fiji would be putting into effect the facilitative dialogue among Parties and non-Party stakeholders that was mandated by PA. Given the sharp swing to subnationals, specifically states, cities, and municipalities, in the aftermath of the realization that while the United States might be not involved in the PA, its subgroups exercised considerable independence, as evidenced by the We Are Still In Coalition (Watts 2017; MacNeil and Paterson 2020), Ambassador Daunivalu underlined the importance of the contributions of subnationals.

Gebru Jember Endalew of Ethiopia, then chair of the LDC group (introduced earlier), welcomed the participants. He commended ecbi on its capacity-building work among junior negotiators, reflecting on his rise from a junior figure to his current position over the course of the past ten years, during which ecbi had supported him. He was followed by Pa Ousman, director of the Country Programming of the Green Climate Fund, who spoke similarly about his rise from the ranks of a junior negotiator of the Gambia to chair of the LDC Group, climate envoy for the Gambia, and now minister of the environment for his country. He spoke forcefully of his indebtedness to ecbi and encouraged the workshop participants to "focus, observe, read and participate actively in the negotiations." While I was touched by his words of care, I was also reminded of Farah Kabir, the Bangladeshi country director of ActionAid, when she introduced me to a young Gambian climate activist, saying, "He has suffered a lot in the hands of his government." The climate process and those who support it must of necessity keep many political tensions at bay in order to keep the negotiations moving, but the suppression of such unpleasant realities sometimes gave a quality of unreality to many nations' statements and actions.

The introductory session was concluded by Achala Abeysinghe, the Sri Lankan–born IIED-based lawyer who was the head of the ecbi negotiator training program. Although her story was undoubtedly also one of going from achievement to achievement, given that she was the legal advisor to the chair of the LDC group, specifically to Endalew, she demurred from speaking of her personal experiences, simply noting the successes of ecbi in having fostered such global leaders as Endalew and Ousman. She drew a link between such capacity-building work and politics by saying that such events and the funding that enabled them helped "resolve differences," "iron out inequalities," and "level the playing field" among developing and developed countries.

At first, I was nonplussed. After all, how did ensuring that a few more delegates than those sponsored by the Secretariat could attend the COPs with some awareness of the issues stand to make a difference within negotiations, given the tremendous structural imbalance among developed and developing countries (Falzon 2021; Chan 2021)? I could see how such events produced the occasion for influencing and networking across developed and developing country lines in the way Mueller had sketched, but I was less convinced by the rhetoric of the "leveling of the playing field."

Through attending and reattending the COPs, I realized the importance of such reiterative engagement, exemplified by Endalew and Ousman. I realized that although country interests might dominate within Parties (Falzon 2021), there was some room for self-modulation through knowing engagement, as opposed to inadvertent cooperation or instinctive resistance to others. The UNFCCC process, directed at producing the sense of a shared planet, if not a shared reality, and a common goal, had a somewhat different air, perhaps of possibility, from other global processes, which were more dogged by inevitability and foregone conclusions. While such capacity-building work by ecbi was didactic and pedagogical and not necessarily directed at enhancing sovereignty or producing solidarity among developing countries, it did have the intent and effect of demystifying the process, to the extent of allowing smaller countries to have some sway within the process, as we will see in what follows.

Learning to Be a Negotiator

After the introductory session, the workshop began in earnest with PowerPoint presentations by the chair of the LDC group; negotiators, including a lead negotiator; and ecbi and IIED experts on what to expect in the upcoming weeks, with a focus on the issues of gender, climate finance, the transparency framework, and facilitative dialogue. This was a mix of topics, with some long familiar to members of the LDC group, such as gender and finance, and some emergent ones, specifically transparency and facilitative dialogue, which were part of the Paris Agreement. There was a midday roundtable in which participants shared their backgrounds, knowledge of the process, and expectations for the COP. But by far the most anticipated session of the event was a workshop led by Dr. Ian Fry, the lead negotiator of the small island nation-state of Tuvalu.

Although Fry was originally from Australia, the fact that he represented Tuvalu within the negotiations and in outside events was not a concern in the same way that a US-based lawyer representing an AOSIS country could indicate undue influence or one for an OPEC country would indicate a bought muscleman. Fry had been invited into the process by the government of Tuvalu and now lived there as a citizen. He had broad-based support across the process, from the Secretariat, to the Bureau of the COP/CMP/CMA, to the blocs of which Tuvalu was a part. One often saw him in the halls of the COP instructing young SIDS-AOSIS negotiators.

Fry was well known for his presentations on breaking down the negotiating process into its constituent elements, with a particular focus on negotiating terminology. He explained what was meant by collective decision making, what a treaty was, and what a decision looked like. In an ecbi presentation to Pacific Island Countries for LDCs, he laid out the sequence of meetings necessary for a decision to be made, starting with (1) meetings held by interested Parties before the COP; (2) bloc meetings of the G-77 and China, AOSIS, AGN, LDCs, EU, and Umbrella group to decide on common positions; (3) the COP opening plenary, in which the agenda for the session was adopted; (4) working groups created by the COP to undertake specific tasks (the Ad Hoc Working Group on the Durban Platform for Enhanced Action that did the work to produce the Paris Agreement was one such working group); (5) informals of contact groups emerging out of working groups to negotiate tricky issues (this was the level at which actual negotiations took place and was generally open to observers); (6) informal informals of contact groups to work more intensely on sticking points (these were neither open to observers nor reported in the daily program); (7) corridor discussions for exchanges of views; and (8) friends-of-the-chair meetings with the chairs of the contact group, to which prominent negotiators were invited to resolve differences among them, with decisions then flowing back up the hierarchy, through the informal group, the contact group, and the working group, to be presented as texts for adoption within the final plenary of the COP. There might even be huddles in different corners of the room during the final plenary to finalize negotiations.

While at first I was astonished by the numbers of meetings that had been wrung out of the process, on reflection what I was most struck by was that the negotiation process had found a way to capture every element of human communication, with formality incorporating degrees of informality and vice versa. More significantly, however, the degrees of formality corresponded to degrees of transparency, with the plenary open to all, contact group informal meetings potentially open to observers unless a Party asked that a meeting be closed, and all informal informals closed to observers, with the minimal transparency requirements met usually by the mere reporting of these meetings over the CCTV.

Fry's description of these meetings confirmed that in addition to following the daily program and the uploading of texts into the UNFCCC website to follow negotiations, one had to be scanning the plenaries, the halls, the meeting rooms, and the corridors for movements and gatherings of key figures to understand the shifting tides of an issue.[5]

Fry's most helpful intervention in his presentations was his sifting through the negotiation terminology to give a sense of how meanings and levels of obligations were changed by the addition, substitution, or elision of words. A sentence might contain all the bits that people wanted within it but might mean something completely contrary to what those people wanted to achieve with it. If the LDCs, for instance, wanted to see a sentence within the text that required that developed countries replenish the various funds under the Framework Convention, such as the Least Developed Countries Fund, they might push for a text such as "Agrees that developed country Parties shall contribute to the Least Developed Countries Fund." There, "shall" indicated a near-legal imperative, but it became a mere ask in the following, simply with the substitution of "invites" for "shall": "Invites developed country Parties to contribute to the Least Developed Countries Fund" (Decision 4/CP 21. National Adaptation Plans).

Fry helpfully provided a list of "slippery negotiating words," with particular focus on the use of "may" (optional requirement at the discretion of the Party), "should" (an obligation that is not compulsory), and "must" (same as "shall," a compulsory obligation). Terms such as "consider" indicated deferral, "organizing a workshop" indicated delay, and "towards" indicated "never getting there."

In his presentation to Pacific Island Countries, Fry ended with eighteen negotiation tips that appeared commonsensical; each of which was accompanied by a cartoon or photographic image that suggested its impossibility. For instance, tip# 1 was "Invest time in knowing the issues you are dealing with" and had a cartoon of a man reading an enormous thousand-page tome, with no possibility of him reaching its end anytime soon. Tip #2, "Hear what others have to say, particularly what they want," showed a cartoon man shouting at the same time that he had his hand cupped behind his ear as if to hear better. Tip #3, "Know the views of your negotiating partners," had a photograph of a man in a suit with his ear pressed to a drinking glass he was holding to the wall, evidently in an effort to spy.

The tips continued and suggested less a political negotiation, in which Party positions and wriggle room were fairly well known, and more of a business negotiation in which there existed the possibility for finding radically new common ground, trading freely, and locking in agreement.[6] Here one has to remember that while negotiators of all shapes and sizes, ages, and experience might negotiate text in the first week of the COP, it was only during the second week of the COP, at what is called its "High Level Segment" attended by high-ranking officials, such as ministers and even heads of state, that is, individuals authorized to adopt decisions on behalf of their nation-states, at which text was further negotiated and adopted as COP decisions. And in the instances of conventions, protocols, and agreements, even the adoption of COP decisions did not commit any one nation-state to putting them into force until the country signed the treaty to ratify it. Furthermore, until a certain number of signatures were garnered, the treaty might not go into effect, further leaving the Party off the hook to act on it. Thus, rather than see the negotiations as a space where positions were genuinely open for negotiation and where commitments once made were relatively firm, it was more appropriate to see them as a space in which Party positions were protected and commitments minimized, with many loopholes through which to extricate oneself from doing anything at all. Far from helping me understand how meaningful and enduring agreement emerged out of this process, Fry's presentation helped me understand how formal and informal procedures and forms and norms of negotiation worked to uphold the status quo.

If anything, Fry's presentation raised for me the question of how to explain any decisive action. A possible explanation was the force of diplomacy itself, that the structure of diplomacy required collective decisions to maintain the sense of interconnectedness of nation-states through the display of tact and sensitivity toward one another. The commitment to diplomacy as the way to transact with one another (although not the only way) meant that the possibility of not doing one's part, and of thus appearing recalcitrant and ungracious, modulated national self-interest.[7]

In the exercises that Fry ran with the LDC representatives at the ecbi workshop I attended in November 2017, he provided four bits of text, asking the participants how they would defer the outcome of each of them. Two were from developed country perspectives, one from a developing country perspective, and one from a country with an indigenous population. He used the same example as earlier on, about how a developed country could avoid having to pay into any funds under the FCCC. The text under negotiation was "Agrees that the Adaptation Fund *shall* serve the Paris Agreement." The prompt to the participants was: "If I was a developed country Party and didn't want a conclusion on whether or not the Adaptation Fund would serve the Paris Agreement because I wanted this decision to be held pending until other issues were resolved, how could I change this sentence?" And quickly the text became: "Recognizes that the Adaptation Fund *may* serve the Paris Agreement, subject to the relevant decisions by the Conference of Parties serving as the meeting of Parties to the Kyoto Protocol and the Conference of Parties serving as the meeting of Parties to the Paris Agreement" (my emphasis).

In the case of the text to be changed to the benefit of a developing country, the text was as follows: "Agrees that information to be provided by Parties communicating their nationally determined contributions, in order to facilitate clarity, transparency and understanding, *shall* include quantifiable information on the reference point time frames and/or periods of implementation . . ." The participants were asked: "If you are a developing country who didn't want the criteria for preparing NDCs to be too prescriptive, how could you change the sentence?" The final text read as follows: "Agrees that information to be provided by Parties communicating their nationally determined contributions, in order to facilitate clarity, transparency and understanding *may* include, as appropriate, inter alia, quantifiable information on the reference point (including, as appropriate, a base year) time frames and/or periods for implementation."

These examples underline my contention that deferral rather than decisive action was the mode of the negotiations. Fry's intervention into this field of interaction was not to give negotiators from Pacific Island Countries or LDCs undue hope about the outcome of their work or the larger process. It was to demystify the process. It was to make them more aware of the nuances of terminology. It was to make them mindful of tricks. And, most paradoxically, it was to teach them how to slow the process. But what Fry taught was not merely deferral for deferral's sake. The negotiators needed to be able to recognize others' attempts at deferral and to concede that sometimes these attempts were appropriate, such as when the final decision on the Adaptation Fund ought to await the final decisions on other parts of the Paris Agreement; to realize some were tricks, such as the call for workshops or the idea of working toward something; and to understand that others were necessary moves on behalf of one's own Party so that one did not commit to something one was ultimately incapable of putting into effect. And, from the perspective of the norms of diplomacy, tip #17, "Don't give in too early," suggested a slowing of the inevitable to enable other factors, such as reading one another's desires and intentions, shaming, or introducing more wriggle room, to emerge so as to bring about more action than any swift course to an agreement.

In their introduction to *Climate of Injustice: Global Inequality, North-South Politics, and Climate Policy* (2007), J. Timmons Roberts and Bradley C. Parks call these modes of stalling and deferring "non-cooperative behavior," making the interesting claim that although developed countries often backpedaled on their promises within the climate negotiation process, noncooperative behavior was more evident among the developing countries. They explain it by showing how structural inequalities in other domains of international relations, such as trade and finance, marred existing relations between North and South. As a result, developed countries were met with distrust and noncooperation in the domains in which developing countries had equal say. While we might be led to think that the climate process, therefore, merely served as a pressure point to negotiate better terms and conditions in other domains, many also claimed that this was one of the few global arenas in which historically weaker countries could participate in meaningful ways.

The Reach of Experts

The ecbi-run day-long workshop came to end with a rousing dinner with boisterous speeches, much back slapping, and a group photograph in which I was included in my capacity as a civil society observer. I went home with reports and flyers of upcoming events provided by ecbi. Looking through them in the evening, I was struck by how smoothly IIED and ecbi made their sponsors known through their logos. It was the widespread use of logos in all COP-related paraphernalia that made me first think of it as a joint state-corporate-sponsored world's fair.

And it was through ecbi's material that I realized that the UK government's Department for International Development (DFID) had funded this project. At one level, this was not surprising because, as Benito Mueller had already indicated, ecbi was conceived to enable Europe to take leadership of the climate negotiations with the help of developing countries. It was a state-oriented project to increase negotiation capacity within developing countries, so funding from DFID was appropriate. However, the incentive structure within the negotiation, whereby DFID would help pay for LDCs' negotiation capacity, struck me as paradoxical, even perverse.

Climate change was a problem created by industrialization. If industrialized countries had taken it upon themselves to mitigate among themselves without involving developing countries, no one would have protested. Instead, under the terms of the Convention it was to be mitigated with every country's involvement. That meant the diversion of money from mitigation to building up the capacity of developing countries, particularly LDCs, so that they might properly participate in the negotiation process. And, as Fry's presentation suggested, what countries did in the process was seek ways to leverage their individual and collective interests, which at many points meant further deferral of mitigation by all.

While training in negotiation produced capacity within developing countries, particularly LDCs, to both hasten and slow down the process so that more ambitious action regarding climate change could be taken, their participation in negotiation was also to serve to spur developed countries to do what they ought to have done in the first place. The negotiation process seemed like an elaborate way to get industrialized countries both off the hook *and* to do what they ought to have done at the outset. In effect, industrialized countries effectively funded others to defer action as well as to goad them into action.

Such insights still would not fully explain the entanglements among Parties, blocs, and the experts who served them.[8] As mentioned earlier, Saleem had brokered a special relationship between the UK-based research organization IIED and the LDC group. That was the reason IIED copartnered with ecbi and why the IIED-based lawyer Achala Abeysinghe came to be the advisor to the head of the LDC group. We already heard from Endalew of Ethiopia and Pa Ousman of Gambia, two key figures within the LDC group, of the years of financial support, training, and hands-on experience they received from IIED to be able to rise to their current positions of leadership within the LDC group. In a separate meeting with Endalew, he underlined that the LDCs had acquired some measure of internal integration and acknowledgment within the negotiation process only thanks to IIED's support.

The LDC group's reliance on IIED was visible. Abeysinghe was always seen with Endalew at the COPs, walking with him from meeting to meeting, sitting next to him on podiums or at plenaries, leading him through agendas of meetings, and socializing in the evenings. Sometimes IIED representatives even represented the LDCs within negotiations, such as the times I ran into Subhi Barakat, another IIED lawyer very involved with the COPs, at bilateral meetings to which he came alone. While it was evident that IIED was sincere in its desire to help the LDCs, that its interest arose out of its mandate to link sustainable development with global climate change, the events that IIED planned at the COPs, such as the weekend-long Development and Climate Days, had a distinctly apolitical feel to them. Its work felt humanitarian, that is, informed by the tendency to do good, and IIED reports, blogging, and public interviews had the unmistakable aura of development expertise. Rather than break down negotiation discourse into common parlance for easier assimilation, IIED discourse met it at its level of involvedness.

After several years of seeing Abeysinghe and Barakat busying themselves around Endalew, I saw a new person similarly busying himself around Endalew at COP24 in Poland in 2018. Manjeet Dhakal was a young man who had served in Nepal's delegation at the COPs since 2009 and who was now the head of the apparently one-person "LDC support team" undertaking "climate diplomacy" on behalf of Climate Analytics, which described itself as a "multidisciplinary and culturally diverse team composed of experts" running the gamut from climate science to finance, adaptation and mitigation, and negotiations.[9] Unlike IIED, which had a wide roster of engagements, from biodiversity to climate change, Climate Analytics seemed a creation of the process, being composed of individuals who had been part of the process as lead authors to IPCC reports or as members of delegations and who now provided their expertise to such figures.

The first time Saleem was at the COPs with a substantial number of his own staff from the International Center for Climate Change and Development, the research organization that he established in Dhaka after he left the IIED, he allowed himself to gloat a little that "his people," that is, his team, equaled the number around Dr. Bill Hare, the head of Climate Analytics. The think tank competed with the IIED to extend its expertise to the LDCs, and its team was always clustered around a table that served as their impromptu office at the COPs. Saleem introduced me to Hare over email, and I began to attempt to meet him or anyone else from Climate Analytics from 2017 onward.

I had heard that Climate Analytics apparently took a nonpolitical approach to climate policy, appearing to put science and evidence-based research, such as the commitment to a 1.5°C rise in temperature, over other considerations. However, their funding from Parties and corporations made their purported neutral approach to climate policy seem dubious. Apparently, they withheld relevant science if they felt that the countries they advised were not supporting appropriate positions, for example, setting aside the Convention to focus exclusively on the Paris Agreement. After hearing all this, it was a bit unnerving to see how close Dhakal was getting to Endalew and how heartily both he and Climate Analytics were thanked, along with Abeysinghe and the IIED, when Endalew stepped down as chair of the LDC group at COP24.

It was a stunning revelation, then, to corner Hare in Katowice at COP24, speak with him for over an hour, and realize that I liked the man, this antiChrist of the negotiation process. A bald, bespectacled Australian, Hare drew me into his world, in which Parties and blocs had long histories and evolving personalities, whose negotiation positions were taken more out of habit than from reason, and whose objectives were obscure even to themselves. I liked him because he was an intellectual who was committed to the grinding work of negotiating but maintained a historical perspective and evinced tremendous self-awareness about how he operated and was viewed by others. While many I spoke with at COPs over the years appeared to enjoy the opportunity to reflect on the process when I asked them to, Hare displayed a far broader scope in his understanding of the process.

Hare was clearly enamored of countries such as the United States, which he described to me as having the necessary gravitas and governmental structure to actualize climate policy and without whom the process was incomplete. He was among those who felt that everything should be put on hold until the United States rejoined the process. Perhaps it was wishful thinking on my part, but I felt that the time had come to imagine a world order in which the United States was not at the center. Hare was not leading us to that day.

Experts from Climate Analytics were as much at front and center as those from other expert and advocacy groups. They often sat beside country negotiators and huddled with them when needed to make an intervention. In addition to the LDC group, Climate Analytics also worked with certain countries within SIDS-AOSIS. It was related to me that certain Parties in the AOSIS group had been the ones chastised by presiding officers for appearing to mindlessly mouth the positions of their foreign experts, who provided them support but also likely used them as mouthpieces. Climate Analytics appeared to have built up a negative reputation for itself for doing so. On the day I met him, Hare related that a female member of his team had been shouted at by delegates, who suspected her of overreaching in her advice, pushing her to tears.

Given the history of SIDS-AOSIS, it should not be a surprise that Climate Analytics had come to advise Pacific Island nations in the first place. Climate Analytics had the reputation for putting science first, and that was the position with which AOSIS led its moral crusade within the process, with science predicting the demise of low-lying island nations and the need for the world to step up. AOSIS's science-led position was also not surprising, as it itself was the creation of US- and UK-based lawyers who thought to use island nations as the means to draw global attention and produce urgency around climate change in the early 1990s (Betzold 2010).

Not all groups seeking to offer their expertise to developing countries took such a hands-on approach. Islands First, a New York–based organization, also worked with SIDS-AOSIS. However, as Mark Jariabka, a US-based lawyer within Islands First explained, his organization had tried to develop a philosophy of expert advising that departed from other experts on ethical grounds. It merely offered its expertise to the AOSIS Parties so that they could develop common positions among themselves to put up a united front within the negotiations. And toward this end, Islands First offered its skills in reading and analyzing documents and writing legal briefs and policy positions but refused to advocate for any position or to speak on behalf of Parties or in place of delegates within sessions.

Experts ran the gamut from large organizations, to specialized organizations, to individuals. In the few but notable instances I have been describing of figures providing expert guidance to countries in the Global South, it should be clear that providing expertise included funding, training, researching, representing, advocating, or helping with the procedural and technical details of the process, or some combination thereof. Experts exhibited various degrees of visibility, with IIED and Climate Analytics front and center and Mueller's Oxford Policy Group, which was part of ecbi, and Islands First very much behind the scenes. They espoused different political orientations to providing expertise, from developmental pedagogy, to scientism, to producing mutual understanding across the developed and developing country divide, to encouraging self-reliance.

One day in the corridor in the COP meeting halls in Katowice, I saw Meena Raman, a Malay Indian lawyer and activist who worked as a researcher, as many experts in this field are called, for the Third World Network (TWN) and served as head legal advisor for the LMDC group within the climate negotiations. Meena was suspicious in manner, snappish to those who she felt were wasting her time (such as myself), but warm and quick to embrace her old comrades (such as Asad). As I stood around shamelessly eavesdropping, she berated the head of Iran's delegation, who was also the spokesperson for the LMDC. A tall, large man, he had to bend over to hear the diminutive Meena, who kept up her invective for ten minutes. Corridor conversations were both important and fair game within the process.

Meena was furious at the delegate for not having followed her instructions, making evident what was widely known, that while the LMDC was autonomous in many ways, it took guidance from the TWN as well as South Center, yet another organization providing expertise within the climate negotiations. On other occasions, I observed Meena sitting right behind the negotiators for the LMDC in negotiation rooms, regularly passing up notes to them from the back and huddling with them at every opportunity. However, as Gary Theseira of Malaysia (introduced earlier), who was lead negotiator within the G-77 and China and spokesman for the LMDC, reminded me, the two organizations were started by countries and people from the Global South, as opposed to the other organizations started and funded by countries in the West. They were homegrown experts, somewhat akin to the kind that Saleem was attempting to grow in Bangladesh, although they were allergic to foreign funding in a way that the ICCCAD in Dhaka was not.

In contrast to Meena's in-your-face presence within the negotiations, there were those who worked as sole agents in the shadows. I have been asked not to identify them by name in my work. One was a lawyer favored by AGN, who wrote statements, read negotiating texts, and introduced new text, all the while moving smoothly from individual to individual, weaving them together into a wider network with an eye toward political change. The questions that guided his actions were: "What is the theory of social change here? What am I trying to achieve? How can it be achieved structurally, rather than through interpersonal relations alone?" Another was a college professor who also worked with AGN behind the scenes to help develop their position on issues such as loss and damage. It may be that the AGN preferred to keep its occasional use of foreign experts within negotiations undisclosed, given its preference for homegrown experts. And for the experts, anonymity was preferable, because the thrill lay not in publicity but in catalyzing change. While I might sympathize with them, I realized that if they could operate so effectively in the shadows to produce the kind of politics resonant with mine, there must be others in greater numbers operating in similar ways to engender a different kind of politics.

4

Politics in Between-Spaces

At the training for developing country negotiators ahead of COP23 in Bonn, I found myself sitting next to Dr. Ian Fry at lunch, the Australian professor of international environmental law who had been representing the Tuvalu government in international contexts, such as the UNFCCC negotiations (introduced in the previous chapter). Fry had just taken participants at the training workshop through the ABCs of effective negotiation. After introducing myself, I asked if I could speak with him about his experiences. After initially appearing to agree, which was a nice surprise, since people in the climate negotiation process tend to be at their busiest during the COPs, Fry politely declined, saying he did not want to focus attention on himself as an individual. The process was much larger than any one person, and it took the labor of many to accomplish anything. His explanation resonated with me.

One of the challenges in writing this book has been to convey the magnitude of the climate negotiation process and how this is both a strength and a weakness. From the outset, it was the individuals in the process who pulled me in. I was inspired by their descriptions of their work as a calling. It took many encounters and conversations with the people who appear in this book for me to see the process as both structured by history and geopolitics and inflected by deep individual commitments and interpersonal relations.

In this chapter, I switch from discussing the process and its various parts to providing portraits of five individuals within the process, discussing their work and motivations for being in this process, before taking us into the thick of the technicalities of the Paris Agreement and the negotiations surrounding it. In doing so, I realize the truth in Fry's words. It wasn't Obama speaking to Li Keqiang in 2015 that guaranteed that China would agree to the Paris

Agreement. It was a Canadian-born official at the Secretariat who ensured that everything was well planned for the meetings; a Bangladeshi scientist who made it his life goal to bring LDCs to the negotiating table as equal partners; an American Quaker who brought delegates from diverse Parties together for dinner so that they might know one another as human beings; a Filipina activist, official delegate, and policy wonk; and a French lawyer who worked furiously to bring diverse civil society constituencies together speaking with one voice.

"Painting a Bull's Eye upon One's Chest"

I decided to speak to someone in the UNFCCC Secretariat belatedly, after my fieldwork was almost complete. Johanna Depledge (2013), a scholar of the negotiation process, has written that if the Secretariat was invisible to those attending the COPs, then it had done its job well, because it was there only to facilitate the meetings, not lead or guide them. I had from the start noticed and admired the infrastructure supporting the meetings, from the particle-board frames of the temporary meeting structures; to the security arrangements; to the booths at which to collect one's IDs, goody bags, and negotiation texts; to the coatrooms, tea shops, and cafeterias; to the smiling young volunteers handing out bars of chocolate and providing directions and the more official-looking men and women with blue badges sitting alongside the presiding officers at plenaries, meeting rooms, and Secretariat-sponsored side events. The most public face of the Secretariat was its head. Christiana Figueres was the executive secretary when I first started attending the COPs in 2015; the position is currently held by Patricia Espinosa. The Secretariat was omnipresent yet invisible: Meeting participants took it for granted, and Secretariat staff did not speak to the public.

But as time went on, I started to notice that the Secretariat wasn't simply fulfilling its mandate of facilitating negotiations but also steering the process. The issues covered at side events weren't just those mandated by Parties at prior meetings but also provided education on emergent issues, such as the circular economy, and expert training on longstanding issues, such as carbon inventories within nationally determined contributions (NDCs).

And activist friends groused that progress in the process had long been held back by the Secretariat's early investment in emissions reduction. By developing staff capacity for complex carbon inventories and reviews, the Secretariat was bound to this course over others. It made it such that the Secretariat now was loath to go beyond "bean counting" at a gross national level to undertake the complex job of bringing large sectors that fell outside of clear national jurisdictions, such as shipping and air freight, under appropriate and more granular emission reduction guidelines.

In my own research into the negotiations of texts, I noticed just how much work was done to synthesize competing positions within a shared document through neutral-sounding tasks, such as "compiling," "reducing redundancy," producing a "navigation tool," and so on. This work, done by the Secretariat at the behest of the Parties, which carried the legitimacy provided by the presiding officers (Depledge 2007), clearly had an outcome in mind, even if just the limited one of moving the process along. So how was I to speak to someone in the Secretariat to get its perspective on the process and its role within it?

A fellow anthropologist studying the process from within the Secretariat introduced me to Richard Kinley, who served at the UNFCCC between 1993 and 2017 and as deputy executive secretary for his last eleven years there. He was retired now and free to talk with me. In tribute to his service, the UNFCCC had produced a Richard Kinley Gallery at its headquarters in Bonn, which recounted the history of the process. I had perused it over the course of attending intersessional and COP sessions. It felt a bit like I was meeting living history.

Kinley and I met and spoke via Skype three times in May 2019.[1] Our conversations were wide ranging, spanning his educational training, early years working for the Canadian government as part of its climate delegation, and twenty years plus working for the Secretariat, coupled with his reflections on the COP meetings that produced the Kyoto Protocol and the Paris Agreement and comments on the current state of affairs regarding global climate change. Kinley was open, frank, and evaluative, even critical at times of what was happening with respect to climate change. However, I was also hearing something that he had polished and refined through repetition, and I wasn't surprised to hear a more recent interview with him on YouTube where he voiced similar critiques with the same turns of phrase he had used with me.[2]

No doubt all of us are written over in some way or another by our upbringing, class background, education, and vocation and that over time we become more set in our speech and gestures. But the way an inveterate diplomat sounds is distinctive; it's a mode of speaking without giving offense. Kinley personified this. Listening to his words one was treated to a kind and polite man with politics that were progressive with respect to climate change. He offered a clear indictment of our present in failing to do what it must to protect humanity, but always couched in inoffensive phrasing. What I attend to in my interview notes is when Kinley spoke of being "in-between," sometimes exiled from safe shores, and sometimes in the flow of things. This perspective helps me draw out some aspects of being a bureaucrat in this very particular UN institution.

Kinley majored in international relations for his bachelors and masters. From 1990 to 1993, he was part of the Canadian delegation to the COPs. He was drawn to the work not because of its focus on the environment but because it was international. Talking with him, one senses an attraction to a world beyond his own: He went to the United States for his education and was drawn to international diplomacy. When asked to join the Secretariat in 1993, Kinley agreed, though he thought he would likely return to working for the Canadian government after some time. However, he recounted, he underwent such "a psychological shift" in moving from the position of a Canadian diplomat to working for the United Nations that he never again entertained the possibility of returning home. He worked for the UNFCCC for over twenty years, and after he retired, he stayed on in Bonn, where the Secretariat is located.

Kinley described the shift that occurred in him in a somewhat joking manner. He took the oath of office and with that assumed the UN's particular emphasis on integrity and ethics, which was tantamount to "painting a bull's eye upon one's chest." Here on out, he had to "sit and take" all criticisms leveled at him or his institution by governments of the world's nation-states, because national sovereignty took precedence over all other principles within the UN. But he was careful to add that this position of "political neutrality" did not imply "impartiality," because the UNFCCC had been created for a particular purpose, which was to solve the problem of climate change. Every action that the Secretariat took both had to respect sovereignty and forward the cause of climate change. Depledge (2013) describes this as serving two masters, and Kinley described it as akin to "walking a tightrope." Despite having become an UN official, he was encouraged to stay in close touch with the country of his origin because the UN counted on its servants to cultivate and draw on these relations, thereby keeping it informed and influential through diverse channels.

The early years of the Secretariat were rocky precisely because the structure had to accommodate both UN officials and political appointees. Nor was the UNFCCC like other UN organizations. Although under the rubric of the larger UN, with its executive secretary appointed by the Secretary General of the UN, it served only the treaty and had its own funding structure. It was seen as a technical outfit tasked with a time-bound problem. But, Kinley explained, being in Bonn away from the UN in Geneva and under the skillful direction of Michael Zammit Cutajar, the UNFCCC's first executive secretary, helped establish UNFCCC's independence and mode of running things.

Kinley tried to provide a sense of the steep challenges faced by the Secretariat in the early years. "You couldn't even use the word 'carbon' because it went against the national interests of many," he declaimed. "In those days it was all about the economy and the North-South divide." But one had to just "keep going." It was a testimony to the albeit limited success of the process that now "everyone thinks of the climate as a global heritage."

In 2006, Kinley became deputy executive secretary of the UNFCCC, putting him in charge of running all major meetings within the climate negotiation process. In an interview he provided for the Profiles of Paris website,[3] which archives memories of COP21, Kinley says, "I often liken a COP to an ocean liner with a huge number of moving parts, plus crew and passengers, all of which need to function seamlessly if the ship is to reach its destination port and navigate through treacherous waters." And he told me that while pulling off a COP was an achievement, the transition from one country Presidency to another was perhaps the trickiest operation he had to oversee.

He was involved in organizing the COP where the Kyoto Protocol was decided, and he was at the helm of the COP at which the Paris Agreement was reached. While proud of both achievements, Kinley reflected that the UN's earlier success in dealing with the ozone problem through the Montreal Protocol had shaped the KP and that it took the UNFCCC some time to realize that "the climate was not like ozone . . . it was a far more complicated thing." I took him to mean that while it was fine to treat ozone as a technical problem to be addressed through a clearly worded agreement with a strict monitoring, review, and verification process, a highly heterogeneous problem like climate change was not amenable to a one-size-fits-all, overly technical approach, as it had been framed in the Kyoto Protocol. Climate would require much more finessing of countries for a global agreement.

Kinley related that this understanding of climate as a political problem was compounded by the failure of the COP in Copenhagen, which ended with no decision (see Marsden 2011). I had heard many analyses of this failure, most notably that the Danish COP Presidency had acted in a high-handed manner, coming to the meeting with a text already in hand to be negotiated by the developed countries, to the exclusion of the developing countries. That text had produced deep distrust of the process, Kinley confirmed. But he added an additional detail I hadn't heard before. Unregulated access to the meetings by civil society also contributed to the disarray in the negotiations.

Henceforth, the Secretariat would be involved at every step in writing the texts, with the presiding officers of the negotiations extending their legitimacy to the texts. In Kinley's words, "Delegates speak of countries having to control the text. This isn't how it works. There can be no progress this way. The Secretariat prepares the text in consultation with the Presidency." And the Secretariat would manage civil society participation in the negotiations. The daily distribution of a limited number of slips for the civil society constituency groups that I mention in the first chapter was a result of the experience in Copenhagen.

"With the Paris Agreement, the international process had finally delivered on its promise," said Kinley. Along the way came the realization that the United States and China were the countries most necessary to bring on board. With the United States, one had to "learn to go with the flow" because it switched positions so often, going from being involved, to leaving agreements, to acting as a spectator. Whereas with China, one had to tarry with its own presumption that it was going to stay a developing country forever, a position that produced no end of tension within the G-77 bloc to which China belonged. Kinley commiserated with the leaders of G-77 that they had a hard job to do to create consensus across diverse constituents.[4]

Kinley said he was astounded when the text of the Paris Agreement was unveiled, as it exhibited a commitment to inclusivity and transparency he had never seen before in a negotiated text. And he considered it a "miracle" of sorts that the agreement still had a clear goal, specifically to keep temperature rise under 2°C. And I took note when Kinley insisted that "the real story is NDCs, emissions, adaptation funding. Everything else is noise." These were clearly the ways that the Paris Agreement was going to be operationalized.

Kinley said that he was very disappointed that emissions had continued to rise even after the agreement was signed in 2015. He considered this to be a "lost decade," in which the "threshold of leadership had not yet been met." As a reminder to the reader, the UNFCCC process early on privileged leadership over naming and shaming as the way forward. Kinley's analysis was very much in line with the UNFCCC's framework. Although he retired from the Secretariat in 2017, he continued to be involved by giving public talks and interviews and participating in projects such as the Profiles of Paris website. One had to keep going, echoing what he said earlier. It was a politics of optimism.

"A Wearer of Many Hats"

Dr. Saleemul Haq, Saleem to almost all who know him, young and old, was a Bangladeshi biologist-turned-activist who introduced me to the global climate negotiations. While he appears in some way in each of the chapters, I decided to provide a profile of him here, to pull together his various elements.

Saleem's role in my research was not unlike the role he played in the 2017 documentary *Guardians of the Earth* by Filip Antoni Malinowski. The film

dramatically recounts the years, months, weeks, days, and even minutes leading up to the signing of the Paris Agreement in 2015. Although Saleem was
initially meant to be one of the individuals followed and profiled in the documentary, over the course of editing the film, Malinowski cast him in the role
of the narrator, allowing him to direct attention to details that would have
otherwise escaped moviegoers looking at images of meeting rooms filled with
UN officials and diplomats.

Watching the film, I trusted what I was being shown most likely because of
Saleem's avuncular presence and authoritative voice guiding viewers through
one of the largest and most complex processes within the UN. It was this initial sense of Saleem as a narrator of the process that led me in an early paper
to describe him as a "climate diplomat" (Edwards 2010), one who mediates climate science and the negotiation process to a wider audience.

Yet to see him only as such would be to miss that Saleem not only knew
and informed about the process to outsiders but that he was a creature of the
process, always in the midst, pulling strands together while letting other strands
loosen. When he narrated, he didn't do so as a learned outsider but as someone attempting to influence the conversation—and who often succeeded in
doing so. For instance, many have credited Saleem with shaping the discourse
around adaptation within the negotiation process, which now is neck and neck
in significance with mitigation. His influence is even more astounding given
that Saleem did not have any formal position within the process: He did not
aid the Secretariat, nor was he part of a government delegation, nor did he
advise any Parties or blocs.

The Demand Climate Justice activists also spoke about shaping the discourse, but more along the lines of needing to "change the narrative," "level
the playing field," and bring about "system change." They pointed to the unequal economic and political system that existed within the global world order and was perpetuated within the UNFCCC process. But Saleem was not
aligned with them when he spoke about shaping the discourse. While he sometimes spoke their language, such as when he called for divestment from fossil
fuels or a tax on polluters, by and large he felt these activists espoused a Third
Worldism that would not get the Global South very far.

Nor did Saleem align with states, for instance, when they called for a mutual respect of "national sovereignty," insisted that this was a "Party-led process," or were vigilant that the text reflect their position so that they could claim
"ownership" of it. Saleem often told me that such fussing over the wording of
texts was best left up to the Parties, while civil society, including himself, pursued more realizable goals, such as ensuring that adaptation was taken up at

the national level, that there was a serious investment in renewable energy sources, and so on.

So how are we to understand someone like Saleem, who was neither a bureaucrat nor a diplomat, neither activist nor expert, although he dabbled in all four spheres at different points in his life? Marilyn Avrill, a law professor based in the United States and an old friend of Saleem's, who ran the RINGO constituency for many years, provided an apt description. "He is a COP entrepreneur. He sees that there is a need for something to be done. He steps up to do it. And he makes an opportunity out of it," she told me in Katowice in 2018 as we grabbed a cup of coffee and chatted about the various figures in the process in between her many meetings.

Things fell into place when Marilyn made her observation; I realized one could see Saleem's life trajectory through this prism. After his graduate training in plant sciences, Saleem started teaching, only to realize that Bangladesh lacked an environmental movement and advocacy group. He started the Bangladesh Center for Advanced Studies with Dr. Atiq Rahman (introduced earlier), a fellow scientist, in the late 1980s. It was one of the groups that strongly protested the Ershad military government's approach to floods in Bangladesh. Saleem, Atiq, and others argued at the time that straitjacketing the country's rivers pathologized floods; the correct approach was to live with the floods, which were necessary for a delta.

The fight against the Flood Action Plan, which had been extensively funded by foreign aid and managed by the World Bank, showed Saleem another problem—the people of Bangladesh could not determine their own fate because they lacked a seat at the table. He left Bangladesh for England to teach there and in 2001 joined the London-based research organization International Institute for Environment and Development, whose mission was to "build a fairer, more sustainable world, using evidence, action and influence, working in partnership with others." Saleem extended IIED's expert help to the LDCs to develop their negotiating capacity and positions on issues such as adaptation, finance, and, later, loss and damage.

In an interview, Endalew, head of the Ethiopian delegation and until recently head of the LDC group on behalf of Ethiopia, recounted how he little understood what was at stake within the process until he met Saleem, who helped him grow and take leadership. Testimonials of Saleem's support were commonplace across the process. So were charges that Saleem served the interests of European nations who effectively empowered him to "peel off countries that were aid addicted" from the G-77 and China bloc. It was easier for industrialized nations to carry the burden of forty-seven countries within the LDCs than to be answerable to the 134 countries within the G-77 and China.

The Climate Vulnerable Forum was another variation of the LDC group that was diplomatically assembled with Saleem's help. Depending on your position, it could either be seen as a further cut into the G-77 and China's internal cohesiveness or as the group that took up AOSIS's call to have temperature change be limited to 1.5°C in the face of Chinese opposition. The political nature of the climate negotiation process finds expression in the ambiguities that shroud such actions, groups, and individuals as Saleem.

At the same time, Saleem was also part of the Intergovernmental Panel on Climate Change, serving as a lead author for several reports by Working Group II on Impacts, Vulnerability, and Adaptation. Here too he saw a gap that he could transform into an opportunity. Saleem realized that while Bangladesh was clearly in the eye of the climate storm and much written about and discussed, there was hardly any serious scholarship from the country making its way into the international reports upon which global policy was based. He returned to Bangladesh in 2009 to set up a think tank called the International Center for Climate Change and Development (ICCCAD).

Starting with a few employees, office space for hosting meetings and seminars, and promises from university friends and colleagues to teach courses, Saleem set about bringing people to Bangladesh to be educated on climate change. ICCCAD was now established; had a roster of research, writing, and publishing projects; and routinely sent a delegation to the COP meetings to provide support to individual countries within the LDCs and cover the meetings for news media back home.

Among the initiatives Saleem started to make Bangladesh the place people go to "to learn about climate change and to see adaptation in action" was an annual regional meeting to showcase community-based adaptations from the Global South. He hosted the meeting for a few years, and it had since become an event that rotates from country to country. Saleem told me, "I am a starter of things. I let others take them forward."

In a later chapter, I will explore how Saleem was part of the effort to educate LDCs about loss and damage to encourage them to join forces with AOSIS, AGN, and AILAC in pushing for the inclusion of the issue within the negotiations and later within the actual text of the Paris Agreement. While Saleem followed the issue closely within the process, he also withdrew from it somewhat once it got going, only staying involved enough to anticipate the next direction the issue might take. He realized early on that the reality of having some payout system within the various mechanisms and funds within the UNFCCC in the face of loss and damage was still very much in the future and that actual instances of loss and damage would likely have to be fought out within courts, both national and international. With nation-states as the first line of defense for those communities likely to sustain loss, Bangladesh would have to do more to be prepared. With his relentless approach to the various ministries and departments within the government, such as environment, disaster, finance, planning, and so on, Saleem got down, at least on paper, a national mechanism for loss and damage and a scheme whereby climate refugees, who were inevitable, in Saleem's judgment, could be incentivized to go to second- and third-tier cities and towns in Bangladesh rather than the overly congested and failing capital city of Dhaka.

A few years after the signing of the Paris Agreement, when it became clear that most attention was going to mitigation, with some to finance and technology, while adaptation was being given lip service, Saleem realized that no one was focused on the article within the agreement dealing with capacity building. This had always been an interest of his, growing out his desire to make Bangladesh the poster child of climate change innovations. He moved to make the issue his own, and out sprung ambitious networks linking universities and colleges across the Global South to provide training to a new generation of students on the climate process, climate science, adaptation, and so on.

Whenever I came to Dhaka from my field site, I would check in, and, if Saleem was in town, we would meet up to talk. We always started with him giving me a quick rundown of the last few things he had done and what was upcoming. Going over my field notes now, I realize I have pages upon pages detailing Saleem's complex and overlapping world. From 2015, I mostly started meeting him at the COPs, where I often found him sitting in a common space with a table tent in front of him with the words "Saleem's Office" printed on it. At first, I thought it was a joke, a reference to his status as insider-outsider. But then I realized it was also a means by which he carved out a legitimate place for himself, creating a space within the frenzy of the meetings into which he drew people into conversation as needed. After all, he dealt in information, and where better to glean or proffer it but in the flow of things?

As my research took me outside of Saleem's world, I learned that along with his supporters, admirers, and even sycophants, he had his share of critics. To some he was an agent of the Global North; others saw him as seeking only self-aggrandizement. He was a product of the media, his constant tweets, Facebook entries, and YouTube videos indicating as much. While I was happy to accept that Saleem might be self-interested, I found more irritating the claims that because his actions might advertently or inadvertently serve other interests, he was the handmaiden of greater powers. I felt it was wrong to deny that

he could author his own words and actions, as that effectively denied him complexity, a tendency that lurked in all domains of the process, as those in the
Bangladesh delegation had encountered (as had I) in other spaces.

What I found more intellectually worthwhile was drawing out the discourse
that Saleem was attempting to shape. If one looks over his record, one realizes
that he was evolving a third way. He was advocating South-South relations. This
meant something specific for him, not "solidarity," as it did for global activists,
or "bilateral relations," as it did for nation-states, but self-reliance combined
with cooperation. I would call this a neoliberal version of South-South relations. Saleem thoroughly embodied this discourse through his continuous
mentoring, extensive networking, and determination not to traffic in cynicism.
It wasn't quite the politics of optimism that Kinley espoused, which derived
from a faith in humanity in general. Saleem was more present minded, determined to see change already in the works and willing to pursue whatever opportunity was available to materialize it.

"Quiet Diplomacy"

Early on, I noted a religious presence within the climate negotiation process.
I had read with great appreciation Pope Francis's encyclical *Laudato Si*, which
argues for care for Earth as our common home and calls for our ecological
conversion to more sustainable lifestyles (Pope Francis 2015). The Catholic
leader's encyclical was released in September 2015 and aimed to foster a positive outcome at the Paris meeting. I went to several side events at the COPs
where I heard the message sent by Ecumenical Patriarch Bartholomew, the
spiritual leader of Orthodox Christians; a damning speech by Reverend Tufue Molu Lusama of the Tuvalu Christian Church; and a discussion hosted
by the Buddhist Tzu Chi Foundation, among other initiatives by what I learned
to call "faith-based organizations" (see Rollosson 2010; Glaab, Fuchs, and Friederich 2018). Faith-based organizations were an informal constituency that
operated through an Interfaith Liaison Committee recognized by the UN.
They were not among the eight official constituencies.[5]

Over the course of conversations with Father John Brinkman of the Maryk-
noll Fathers and Brothers and Valerie Bernard of the Brahma Kumaris World
Spiritual University, in which neither made a single reference to a higher power
but instead spoke of the "human in nature" and of "empowering the human
spirit," I realized that while faith-based organizations were welcomed to this
space and many were present, they did not deploy the religious references that
usually informed their speech and action. Perhaps it was the resolute secular-
ity of the negotiation space or an agreement to a certain ecumenical or inter-

faith approach that led to their emphasis on peace, justice, and rights and the omission of God, Allah, the Qur'an, Jesus, Buddha, or what have you. For Father Brinkman, the holistic approach of the UN's Sustainable Development Goals was akin to a spiritual mission, whereas for Valerie, human rights expressed the universal within the human. Religion entered by the side door via the language of "spiritualizing the process."[6]

My effort to understand what it meant to spiritualize the process in practical terms led me to the efforts of the quiet yet alert American Lindsey Fielder Cook, who headed the branch of the Quaker United Nations Office (QUNO) in Bonn and represented the group at the negotiations. I had already spotted Lindsey, even before being introduced by Marilyn Avrill. I had seen her at various RINGO meetings, noting her tendency to hang back from discussions, although she would speak at points to inform about developments in different parts of the negotiations and even advertise events that she was organizing despite the obvious disapproval of the conveners that this space not be used for advertising one's events. The impression I had was of her stepping both in and out of view, unlike Kinley, who effaced himself, or Saleem, who kept himself in the public eye.

Lindsey practiced what had come to be called "quiet diplomacy" by the Quakers within the UN, where they worked behind the scenes to bring conflicting Parties to the table and to shared understandings. It was a longstanding practice (Yarrow 1978), and the Quaker United Nations Office also published many reports and policy statements to aid its work (see Elliot and Cooke 2016). Lindsey's own work background included a long history in the field of reconciliation and peace building before joining QUNO. Lindsey informed me that while QUNO had been at Rio in 1992, at which the UNFCCC was forged, it later dropped out of the process. When the organization began receiving inquiries from constituencies in the early 2000s as to what it was doing about climate change, it reinstated climate as an issue under the rubric of "human impacts of climate change" in 2011. Lindsey was hired in 2013 after an international search, and she moved to Bonn with her family to work intensively with the UNFCCC.

On a website hosted by Australian Religious Responses to Climate Change on "Living the Change: Faithful Choices for a Flourishing World," a multifaith global campaign, Lindsey gives a personal account of how she came to be part of the process: "I was so caught up in peace and justice concerns as a humanitarian worker in war zones, that I was not very aware of my relationship with the Earth. But as I began to understand how our lifestyles are destabilizing nature, and how rising temperatures could lead to violent, failed countries where I had worked, I began to understand that climate change (and other environmental crises), and in turn sustainable lifestyles, are a peace and justice concern."[7] She repeated this explanation in a 2016 interview available on the QUNO website, perhaps hinting at a need to justify Quaker involvement with something as technical and long term as the climate negotiation process, far from the immediacy of conflict, war, and suffering.[8]

Lindsey ran side events, produced policy briefs and toolkits so government officials and negotiators had at hand arguments for a more ethical and just approach to climate policy, and provided news briefs on meeting developments to the wider Quaker community. But the primary focus of her attention was planning one or two dinners for fifteen to twenty negotiators drawn from poor- and rich-country delegations during each session, to bring them into conversation over difficult points within the negotiations. I had heard about similar trust-building initiatives across the usual divides (developed/developing, rich/poor) from Dr. Benito Mueller, the Englishman who ran the ecbi (introduced in the previous chapter). Lindsey herself told me of technical dinners hosted by the World Resource Institute.

When I asked Lindsey to clarify her work with respect to these approaches, she immediately put a distance between them. "That work [Mueller's] is large scale, with more money and is more explicitly political. In my work I appeal to the human," she said. She described how it took her a year of diplomacy to get people to agree to attend her dinners and another year to ensure that every delegation would send a member. She had to work hard to ensure that dinner guests followed the rules she set down. They could not talk publicly about the dinners. There had to be a balance between rich and poor countries among those invited. She posed one question to them to which each had to provide an honest response. People could only speak once over the course of the dinner. Raised voices and arguments were to be avoided. The idea was to create a "safe space." At the same time, "conversations within them can get heated, but they are often philosophical," she ruminated.

QUNO hosted its first dinner at COP9 in 2013. There have been about nineteen dinners since that date, and they have now become quite established; Lindsey called them "infamous." She kept the number of invitees low to ensure the events retained the aura of something special.

What are some of the questions that Lindsey has posed to dinner groups over the years? "We ask questions such as, 'What do you understand as fair?'" At our second meeting, she continued to elaborate on the questions she had asked, such as "What do your children think of your work?" "What is your greatest fear in coming into this session?" "What is the name of a person whose life would be affected if we didn't do something now?" Some of the questions were quite leading: "What common positions do I believe we have achieved, and what innovative approach would I suggest for a remaining difficult issue, so our work gives the world a real chance of a 1.5°C temperature rise limit? At Paris she asked, "What actions of solidarity do I think we can take to help transform mistrust?"

Lindsey was very careful not to reveal the identities of her dinner participants, as that would compromise the diplomacy she was attempting to put into effect. However, when she learned that I was from Bangladesh, she suggested that I speak with Dr. Nurul Quadir, at that time the additional secretary at the Ministry of Environment, Forests, and Climate Change, head of the Bangladesh delegation to the UNFCCC, and Zia bhai's boss (introduced earlier), to find out his experiences of the dinners. "I may have had him at a dinner or two in which he was very eloquent." And then she said something that under ordinary conditions would have come across as patronizing, but it was a credit to Lindsey's self-presentation and affect that I accepted it as a simple iteration of the facts: "Bangladesh is such a vulnerable country that having that voice [around the dinner table] made a difference."[9]

Had these dinners helped at all with the negotiations? The question was difficult to answer. Lindsey did not think that she could point to any specific breakthrough at the dinners that subsequently informed the negotiations. "Ours is a very small offering," she demurred. However, couched in general terms on the QUNO website is a humble brag with respect to this process. Given that COP15 in Copenhagen in 2009 was widely viewed to be a failure, it was actions such as the QUNO dinners that helped produce the trust that was necessary in making COP21 in Paris in 2015 a success in terms of producing an agreement. And Lindsey speculated whether a Fijian delegate, who had attended a QUNO dinner, had gotten the inspiration for the Fiji-sponsored Talanoa Dialogue from the QUNO dinners.[10]

Conversations with the Bangladesh delegation helped me realize what the dinners might enable in terms of straight talk. Human rights was a major issue for activists and advocates, and while it was included in the Preamble to the Paris Agreement, they now sought to integrate the issue into the different work streams of the Paris Rulebook. In Bonn in May 2018, I shadowed Shawkat bhai, the member of the Bangladesh delegation following adaptation-related issues. At one point, he informed me of an interesting dinner he had attended the night before where he heard fear expressed by a range of individuals and nationalities that the human rights issue was being used by developed countries to maintain an upper hand. While Shawkat bhai took care to inform me that Bangladesh supported the integration of human rights with climate change, he mused about the many ways the issue had been manipulated in the past, which made him realize that its integration would be resisted in the Paris Rulebook. When this fate came to pass in Katowice later in the year, I wondered to what extent the dinner conversation had anticipated the outcome.

What ultimately did I learn about how the process was spiritualized through such efforts as Lindsey's? She recalled that when she first attended the sessions, she was shocked by the extent of depression, tension, bewilderment, and anger she felt within the negotiation rooms. She took to going around the hallways giving hugs to people. She recounted how people broke down at the dinners and cried because of the pressure they were under. They confessed that they knew this process was not good enough. They grieved. She felt that her work brought "a bit of the human into the process," allowing participants to "be themselves," "to be genuine," to allow them to realize that they "could make a difference."

These may sound like trite formulations, but what they point to is her keen attentiveness to the impact of the impersonality of the process upon individuals, who might have notions of how short they were falling of the action required but who had jobs to do, roles to play, and national interests to watch out for. Lindsey's work was not to render the process in terms of a higher purpose but to spiritualize the effort to do this work, so that the fact that this was a very human struggle could come into view.

"A Piece of Philippines"

I met Tetet Nera-Lauron, my Filipina activist friend whom I have already introduced, in 2017. She seemed the perfect person to profile in this chapter because she had straddled so many roles within the UNFCCC process. By the time I met her, she had been an activist pretty much all her life in the Philippines, transitioned to global activism, served as a member of the Philippines climate delegation for a time, and was on her way to becoming a policy wonk. And her Catholic upbringing also attuned her to the religious organizations involved in the process.

The story of Tetet's involvement with the UNFCCC started with her decision early in life to throw herself into activism on behalf of ordinary people in the Philippines, those who desired an alternative to a debt-fueled economy and social existence. The international service institution IBON International inducted Tetet into this mode of analysis and politics. There she undertook work related to "people's economics."[11]

When I asked Tetet how she got from there to the UNFCCC process, she took me through the other international processes she had been a part of before turning to climate change. She mentioned regional conferences on trade, economic justice, and human rights as early opportunities to share knowledge and learn to do advocacy and campaigning. Her earliest foray into global activism was at the UN World Conference on Women, in Beijing in 1995. There she found herself in a gathering of a scale that she had never encountered before; she spent her time just walking around absorbing the busy activity around her, playing the part of a "development tourist." She was there with a group of prostituted Filipina women who had put together a play to present their plight. The Royal Family of Denmark attended the play. "And *that* was the highpoint of my trip," she recounted, laughing at her young self.

The transition to working on issues related to climate change was easy, she said. "Economic justice, human rights, climate justice, it all flows together." She went from people's economics to the discussion of finance within the process. When Tetet spoke of finance, she spoke of it in terms of the language of historical debt and the need to redress the structural inequality that had been put into place by colonialism in its historical and contemporary forms, the discourse Saleem tagged as Third Worldism.

Tetet's first COP was the infamous Copenhagen meeting, and she was as lost as anyone else entering for the first time. On breaks from manning IBON Foundation's booth in the hall for civil society, she walked around taking in other booths and country pavilions and joined various actions. "I was basically campaigning, putting on actions inside and outside official spaces so that governments would know and feel that they were being watched," presumably to put them under pressure to deliver meaningful action and be accountable. She spelled out the concentrated attention that was, and continues to be, given to actions to make them bold spectacles, to grab the attention and hopefully the conscience of delegates walking past.

Tetet realized that these meetings were incredible occasions to network with civil society representatives the world over; there were very few occasions civil society representatives from under-resourced countries could meet. It is noteworthy that she referred to activists within the COPs exclusively as "civil society organizations," or CSOs. I retain her nomenclature. Even as she got more involved in activism, she recounted the deep divides among the CSOs within the UNFCCC space. On one hand there were those who were "total nerds" and deeply within the negotiation process, such as those in CAN. And there was the other "mafia," headed by none other than Asad and Friends of the Earth, who advocated for a more confrontational and critical approach. They fell under the banner of Climate Justice Now, out of which Demand Climate Justice emerged. These were the groups that mediated between the Secretariat and the Parties and the rest of the CSOs, and everyone fell in line with them. However, Tetet related, the lines had become more blurred of late: The

CAN folks had gotten more activist and those in the other camp much more versed in the minutiae of negotiations. Although her politics were in line with DCJ and she affiliated mostly with them, she did not feel that she could exclusively be a DCJ spokesperson, given her many commitments.

Since the early meetings of CSOs at the COPs, there had been a flourishing of alternative meeting spaces that were independently organized but also marred with contradictions. She attended CSO meetings hosted by Bolivia in 2013 and Venezuela in 2014, both progressive Parties with strong leftist politics. In Venezuela there was much debate and heartache among CSOs as they came to the realization that "Venezuela was a petrostate," which compromised one of their basic political tenets: calling for an end to extractivism and fossil fuels. Tetet said it was a bitter truth: "There was nothing we could do about it without turning our back on one of the few Parties that supported our full agenda," which included food sovereignty, intergenerational equity, gender equality, indigenous issues, and human rights, all of which could only make it into the negotiations if they had buy-in from Parties. Also, CSOs were overly reliant on Parties for their observation of negotiations; that is, if a Party requested that a particular negotiation be closed, it shut out all CSO observers from the room.

It was only at her second COP, in 2010 in Cancun, that Tetet felt that she ought to pay more attention to the actual negotiations going on at the meetings and what they meant for the Philippines. "We activists, we bring attention to issues, we criticize if we feel that the work is not getting done, but we don't have the discipline to get into the details." She started to actively attend negotiations on climate finance.

At that point, Naderev "Yeb" Saño was leading the official Philippines delegation to the COPs. It was under his direction that the Philippines was associated with the LMDC bloc, and he brought a lot of attention to the issues of the Global South, including through his emotional speech and public fasting at COP19 in 2013 in Warsaw after the super-typhoon Haiyan hit the Philippines, which gave momentum to the issue of loss and damage (see Rajamani 2014). Yeb Saño invited Tetet to join the Philippines delegation in COP17 in 2011, in Durban, to follow the issue of climate finance, a duty she fulfilled by attending negotiations on the issue and writing up notes for other members of the delegation. Referring to her note taking, Tetet said, "You type and type. Even if you don't understand, you really see the power discrepancies play out within the negotiations." While she did not have any illusion that "these negotiations will amount to anything," she felt responsible to report developments back to those who could not attend or those who did not yet know what was at stake. She was slowly realizing that "there is a policy wonk within me."

Tetet's affiliation with the Philippines delegation lasted the remainder of
Saño's leadership. He stepped down as commissioner of the Philippines' Climate Change Commission on the grounds that he wanted to join the people
in their fight for climate justice. Tetet stated bluntly that Yeb Saño was sanc-
tioned at COP20 in 2014 in Lima and was asked to resign because the Philippine government felt that he had made the delegation his own platform.

Despite these internal political battles within the Philippine delegation,
Tetet came to have a deep appreciation of the hard work of being a Party delegate. She saw the hours they worked, how much they needed to be both open
and flexible within the negotiation while toeing the country or bloc line. At
the same time, she shrugged philosophically, it was widely understood that the
delegates had to return with deals to their country if they were to keep their
jobs. Tetet said that Saño referred to deal making as "forum shopping." In
law, the phrase usually refers to seeking courts that will provide favorable rul-
ings. I didn't fully understand how deal making was forum shopping. However, when Tetet told me that after Saño left the delegation, the Philippines
shifted its alliance from the more political LMDC to the Climate Vulnerable
Forum (the bloc formed of the poorest, most climate-vulnerable countries,
which was asking for decisive climate action but was also viewed as a creation
of the Global North to weaken the base of the G-77 and China), I wondered
if we could understand this shift as the outcome of forum shopping by wealthy
nations. Tetet felt that the CVF promoted the worst kind of "climate porn," by
projecting images of victimhood and suffering.

When Yeb Saño was removed from the process, Tetet went back to doing
what she had done before, representing IBON International, following and reporting on negotiations, and undertaking actions inside and outside official
spaces. But she retained her connection to him. He took up the charge of
organizing and implementing the Climate Pilgrimages, a Catholic undertaking inspired by Pope Francis's *Laudato Si*. The first one went from Rome to
Paris, bringing ordinary people to COP21 in 2015. Tetet joined Saño for the
last leg of his march to COP24 in 2018 in Katowice as a gift to herself for her
fiftieth birthday.

Tetet's changing positions within the process showed how it mattered that
one attend regularly, first becoming visible within the small inner circles and
then becoming integral by having something useful to offer. She had come
in as a tourist of sorts, just advertising her organization and their work at
the earlier meetings. She was consigned to the edges of the process. She had
come to be a reminder of realities on the ground, but as she got more educated on the negotiation process, she started to write analyses for IBON International for public communication. Of these she said laughingly, "They were

less nerdy and more political than the TWN [Third World Network] newsletters." One of the points that she joined others in demanding was a clear accounting of funds provided for climate change by wealthy nations, because they were double counting their development aid as both humanitarian aid and adaptation funding. Her relentless beat gained her the respect of the more activist CSOs and won her invitations to comment on the process. Once she was in the delegation and had ready access to Yeb Saño, her value rose rapidly. "Philippines had just experienced a hurricane. Saño had delivered his emotional speech to the assembled audience. He was a rock star. Everyone wanted a piece of Philippines. I was there and suddenly invited to all kinds of side events."

She was also invited to side events organized by religious groups and organizations. These organizations often sponsored under-resourced CSO participants from developing countries (Bread for the World sponsored Tetet). She also connected to religious groups through activist politics, given their commitment to the poor, articulated through the ecumenical language of climate justice along with SDGs and human rights. In part, it came from her own Catholic background and appreciation for Pope Francis, who to her was also clearly a "rock star." However, here too as with CSO politics, Party politics, and delegation commitments, she provided a wry assessment of the religious organizations, noting that, thanks to Pope Francis, the Catholics were clearly in the limelight in a manner that the Protestant groups and organizations were not, causing some residual vexation.

Tetet was the one who told me about the cross-constituency coalition that mediated between the Secretariat and CSOs. In particular, it took coordination and a shared agenda among the environmental NGOs, the women and gender constituency, the youth NGOS, trade union NGOs, and the Indigenous people's organizations to agitate for progressive language within the 2015 Paris Agreement, which had led to the Preamble. However, she also noted a tendency within the process to co-opt progressive language and render it meaningless, for example, as what happened to the term "sustainable development." Co-opting language was often accompanied by efforts to buy off constituencies. Case in point, the Polish government's focus on "just transition" in Katowice. "Suddenly the focus wasn't on justice before transition, but rather economic transition before we attend to issues of justice," she complained. There was also a lot of horse trading with the constituencies, with Parties offering to take some parts of the package, such as gender equality, over others, such as human rights.

It was her understanding that the process was always trying to co-opt the CSOs, so a large part of activist work was to stay ahead of this co-option and to keep recasting the discourse. While the core issues remained—"historical responsibility," "making good on pledges," "avoiding conflicts of interest," "finance," and "no false solutions"—CSOs were now putting forward the rhetoric of "climate-positive recovery," "carbon neutrality," and "race to real zero."

Over the course of her long arc of involvement with climate activism, Tetet made a hard turn from following and reporting on negotiations to trying to shape them, which marked her transition into becoming a policy wonk. In 2018, she left IBON and was recruited by the German organization Rosa Luxemburg Stiftung to be their regional representative not just to the COPs but to other UN processes as well, a position effectively created for her. As part of her new job, Tetet now organized side events on issues ranging from loss and damage to climate migration.[12]

In our conversations, the comparison or, rather, the contrast between the Philippines and Bangladesh always came up. I draw out these differences to elucidate how two countries, both poor and living with similar environmental challenges, had such different approaches to them. Most of the representatives of Bangladesh CSOs I'd had the opportunity to meet within the UNFCCC space (with introductions often made by Tetet) were involved in NGO work such as primary education, women's empowerment, microcredit, and social enterprise, all cast in the language of social uplift. Tetet and other Filipinos I encountered at the COPs, such as Lidy Nacpil, approached the situation with a much stronger political analytic, one laying the blame for social ills on international and national policies of structural adjustment and advocating for debt-free development. One heard the difference in Tetet's own comparison of the two. "They are very inward looking," referring to the fact that the Bangladeshis tended to stick together at the meetings, "whereas we look outward."

There was also a sharp contrast between the Philippines of the LMDC and Bangladesh as part of LDCs and more recently the CVF bloc. While finance was a right according to Tetet, most of the Bangladeshi delegates didn't see any grounds for countries to lay claim to the wealth of other countries. All they could see the scope for were concessions in interest rates and flexibilities in payment in loans to enable countries further back in the development trajectory to catch up. Tetet wasn't having it, and she teased her Bangladeshi allies mercilessly for their developmentalist mentality and politics.

"To Look Forward Rather Than to Act Retroactively"

At a meeting of the cross-constituency working group on human rights Tetet took me to in Bonn, a group chuckled over a video that had been posted on Twitter of Sebastien Dyuck, a French attorney working in the Geneva branch

of the Center for International Environmental Law. It was a two-minute capture of his sneaker-clad feet as he rushed from place to place at the intersessional session. They laughed when I asked how I might touch base with him,
indicating what I had already surmised. He was too ferociously busy at the
COPs to be met there.

I tracked Sebastien's activities by participating in several public actions he
organized on how human rights rested on all the different principles coming
together (rights of indigenous peoples, public participation, gender equality
and women's empowerment, food security, just transition for workers and decent work, ecosystem integrity). I attended cross-constituency meetings where
he discussed the extent to which the principles in the Paris Preamble were making their way into the implementation guidelines. I heard him at various press
conferences and side events. And I met with him for short bursts of time here
and there, waiting till I was writing this chapter to reach out to have Zoom
conversations with him.

I was interested to include Sebastien because he undertook several media-
tions between human rights and climate change and between existing treaties, policies, and laws that were already in place in international and national
contexts and what was yet to emerge through the Paris Rulebook. He reminded
me that whatever came out of this process was not sui generis but bound by
previous commitments.[13] Or, as he liked to put it, if this process did its work
comprehensively, then it could deliver on its promises. In conversation with
him, I realized that he also brought in another dimension of in-betweenness,
that between youth and the process.

Sebastien described himself as a typical EU product. Although he was born
and raised in France, his education and professional training occurred in
institutions across Europe. Speaking in his usual soft voice but rapid-fire
style, he provided the names and places of these institutions, of which I caught
Germany and Iceland. He opted to study law because he wanted to work
with the "good guys." He studied both human rights and environmental law
because "environmental pollution" was a big issue in his youth. Although he
never gave up either, over time he gravitated to working in the environmental
domain. His thinking was that human rights was reparatory, coming after the
violations had happened, whereas environmental law could be more future
looking.

He only ever worked for NGOs, never for any state, because he didn't have
the patience for the slow pace or long-term planning required of governmental work. And so it was that he found himself at a job in subarctic Finland at the
time of COP14 in 2008 in Poznan, Poland. Given that he looks to be in his
early thirties at present, that would put him then in his early twenties.

Thinking he could be useful to activists attending the COP to navigate a city with which he was familiar, he took off from Finland to offer his services, only to realize that the activists were as organized as the rest of the process. Not only were they organized; they were busy. The pressure to be organized in time for COP15 in 2009 in Copenhagen, at which a new treaty was to be revealed, was very high.

Chagrined, Sebastien said that he looked around to see how else he could make himself useful and realized that he could bring the COP to the city by bringing attendees out of the meeting space and into Polish classrooms. "At that time, hardly anyone knew about climate change or was doing anything about it," he explained. Such "consciousness-raising work" felt to him a natural extension of his previous work educating youth on anti-Semitism, antifascism, and antiracism, yet another route to climate to those already considered earlier.

This was a defining moment for Sebastien. The organization headed by the activist Bill McKibben, 350.org, was at the COP. There was more youth presence than in previous years. In fact, the European Youth Forum brought him into the official space. And he jumped at the chance to help define youth involvement in the process.

"It was in 2009 in Copenhagen that the youth got their own constituency." He paused to make sure that I understood what this meant. One might think that because there were already a few CSO constituencies in the process, notably BINGOs (business NGOs) and ENGOs, it was par for the course for the youth to have a constituency as well. But the young people present at that time debated for a long time whether they wanted to intervene in the process from within it, to submit to officialdom, because that meant communication going forward would be through one or two persons who would be the constituency's focal point(s).[14] They were worried they would lose their critical edge and be capitulating to what everyone already understood to be a compromised process. How could they continue to "blame the process" if they were within it? But eventually they agreed to join, and Sebastien was there to help them set up this constituency with a steering committee, a mandate, and focal points.

The youth were right to be worried, Sebastien mused. He remembered a time when the young proposed to have a Youth and Future Generations day within the two-week-long session in which all the various tracks within the negotiations would attempt to deal directly and decisively on issues that affected the young. The proposal was initially turned down by the Secretariat, who scoffed that "we don't have themed days within the negotiations." However, once the Secretariat relented, they went full-bore, making every day of the two weeks a themed day to draw attention to various issues, which the youth took to be a cooptation and dilution of the kind of exercise that they had in mind. But, Sebastian mused, the problem with the youth constituency was the lack of institutional memory. After all, the young grow up and leave this stream of organizing. "Now the youth feel flattered when they are asked to speak on events organized by the Secretariat on Youth and Future Generations Day, without knowing this background."

While Sebastien was sympathetic to youth concerns, it was clear that he also liked things to be participatory *and* orderly. He spoke approvingly of the fact that the Secretariat no longer had to go scrambling after a headless movement to get their input. Youth were invited to briefings with the Secretariat. And aside from being given the usual quota of badges to attend negotiations, they were also allowed access to roundtables and technical workshops with delegates, which were previously behind closed doors. And unlike others who felt that establishing constituencies and a badge system reduced CSO impact on the negotiations, he tended to sympathize with the need to bring the more than twenty thousand CSOs into manageable form.

Sebastien recalled how the YOUNGOs (as those who belonged to the youth constituency came to be called) continually evolved creative ways to produce meaningful impact within the official space. At COP16 in 2010 in Cancun, youth blogged and tweeted continuously, turning this into an important means to make the COPs real and approachable for those outside. The following year, at COP17 in Durban, there was the action of "adopting a negotiator"—latching on to delegates from their own home countries to be able to glean information from them and to provide CSO feedback to them. This was hugely successful and had continued.

At COP18 in 2012 in Doha, Sebastien pivoted to the intersection of human rights and environmental law. He wanted to make human rights part of the effort to make climate action ambitious. His concern, along with others, was to ensure that human rights safeguards were built into climate action and that climate action interacted and was integrated with everything else. He noted that a Mexican negotiator had first brought up human rights in Cancun, but, although this reference was noted in the official records, it did not gain traction and was not discussed. Given the competition among the different UN bodies, the UN High Commissioner on Human Rights had no more status within the UNFCCC process than any other CSO, so it was not able to insist on the issue.

Sebastien created an informal, cross-constituency working group on human rights to bring the issue more forcefully within the process. Through discussions in this forum over several years leading up to COP21 in Paris, it was decided that all the constituencies within this working group would push one another's platform and collectively insist upon human rights. Sebastien insisted that there was no effort at forging a consensus, in having a unified voice or even a common script. Rather, "we just did our own things, having decided that we will be fair to the others." The constituency system certainly helped here, he pointed out. He took a moment to relish how much it compounded the message when one spokesperson for the constituencies after another spoke up about the same roster of concerns, "the human rights package," within the plenaries. It brought visibility to the issue, and its success lay in "having a light touch" and "relying upon the intelligence of the network." Since that time, Sebastien mentioned "capacity-building sessions" in 2017 on how to take advantage of various entry points, pitch issues, and teach about them as occasions arose.

This collective strategy worked to bring the package within the Paris Agreement, albeit restricted to the Preamble. Sebastien was rarely demoralized; he recast as opportunities what others viewed as failures. Not having the various components of human rights within the articles was a loss. It indicated that human rights needed a higher level of consensus, a high bar within the process, and a higher level of mandate within the UNFCCC. But it was a win from a legal perspective that these rights were recognized. From this perspective, having the human rights components within the Preamble was all that was needed to remind and make actionable all preexisting and relevant agreements within national contexts.

Unfortunately, the Preamble was set aside as negotiations continued beyond the Paris Agreement to its implementation "guidelines," with Sebastien now telling me that negotiators preferred this phraseology over "rulebook": The latter sounded too binding. This meant that one had to attempt to enter the package into every negotiation track under the Ad Hoc Working Group on the Paris Agreement, specifically those relating to mitigation, adaptation, transparency, and global stocktaking, where human rights would have most relevance. While most of the official negotiations emphasized quantification, that is, how much climate action was to be done, human rights advocates (Sebastien saw himself much more an advocate than an activist) emphasized quality, that is, how climate action was to be done.

While the constituencies still insisted upon all seven elements of human rights, it was clear that delegates wanted to cherry pick only those that suited them if they wanted to engage them at all. Most shied away. Sebastien conjectured that much of the resistance to human rights stemmed from fear across the board that litigation was on its way. Mobilization around human rights failed so abysmally in COP24 in Katowice that he started to lose hope that even shadows of references to any of the principles would show up in final texts.

While Sebastien's fears were realized, he still found something to commend within these failed efforts. Of the two objectives, to get explicit reference to human rights within the guidelines and to educate civil society and governments, they had clearly succeeded in the latter. Furthermore, financial institutions and the private sector had received an education in the fact that "harm harmed credibility." He had forged an extensive network through his years of coming to the COPs and had actively brought delegates together with civil society through arranging dinner get-togethers. He knew of Lindsey's work and respected it but saw his dinners as more advocacy oriented.

At COP24 in Madrid, he was heartened to find that half of the world's governments acknowledged the need for real and meaningful emissions reductions to uphold the integrity of the PA and the need for some human rights safeguards somewhere to ensure that the reductions did not do more harm than good. I suddenly realized anew that these advocacy efforts were not only to use climate change mitigation as the way to address the world's preexisting problems, as was the case with Asad and Tetet, but also to protect the world from the problems inevitably to be created by putting such mitigation into effect. Interestingly, the Parties that sought such safeguards were those in the Global North. They were worried about how a new era of interrelating with other Parties, with authoritarian governments or ones with records of mistreating their own populations, through a renewed carbon market to be put into place by Article 6 of the Paris Agreement could entangle them in compromising situations. This was a perspective on the Parties of the Global North I had not encountered before and could appreciate.

Ultimately, Sebastien said, the UNFCCC was a failed process, one that had been failing for a while. All he was doing along with others was bringing human rights to the process, whereas in the rest of the year he attempted to bring the process to human rights conversations, which was equally difficult, if not more so, given the widespread lack of concern with the environment. He saw himself as trying to level the playing field, produce points of pressure, build coalitions, and draw attention to the national spheres where most of the action was. When I asked him about the legal status of the issue of loss and damage, he said that issue represented human rights as reparative action coming after damage had been done. He saw loss and damage as entering this process to redress the failure of bringing in human rights, of potentially anticipating and averting damage.

The five individuals outlined in this chapter exemplified climate politics for me. Each entered the process via a distinct pathway. There was an in-between quality to each of them, Kinley for being between the Secretariat and Parties; Saleem for being between the official and the unofficial dimensions of Parties; Lindsey for mediating between Parties of the Global North and the Global South; Tetet for shuttling between international activist, delegate, and policy work while being rooted in her national context; and Sebastien for working between youth and the process and between human rights and climate change. While the process was gigantic, these individuals, their diverse motivations and means of entry into this space, and their in-betweenness provided important perspectives on how this process branched beyond structure, hierarchy, procedure, and texts.

5

Accounting for Change in the Paris Agreement

At the midyear meeting of the COP, called the Intersessional, in Bonn in 2018, Hafijul Islam, a lawyer friend who was part of the Bangladeshi delegation, told me he missed the elegance and clarity of the Kyoto Protocol (KP). It was two years after the ratification of the Paris Agreement (PA), and the fate of the KP was dire. Although negotiated in 1997, it took ten years for the protocol to get the signatures of a minimum of 144 country Parties for it to be ratified, so 2008 through 2012 served as its first commitment period. While the KP was extended in 2012 until 2020, the second commitment period had yet to go into effect because only 124 Parties had signed it. With only a few years to go until 2020, there was no doubt that the KP would cease to exist after that year. Meanwhile, the meeting of the CMP, the Conference of Parties serving as the Meeting of Parties to the KP, kept congregating under the COP, spinning its wheels until its impending demise.

At the same time, the Ad Hoc Working Group on the Durban Platform for Enhanced Action (ADP), created at COP17, raced ahead to negotiate and draft the Paris Agreement. Unlike the KP, this agreement was ratified a short year after it was negotiated. Once the agreement was ratified, ADP, the group responsible for birthing it, came to an end, and another, the Ad Hoc Working Group on the Paris Agreement (APA), began its institutional life to produce the Rulebook by means of which the agreement was going to be implemented by the Parties. While Hafij labored with thousands of others to determine what the agreement would look like in practice, the CMA, the Conference of the Parties serving as the Meeting of the Parties to the Paris Agreement, which had started in 2015, continued meeting without closure until such point as the Paris Rulebook was completed and adopted in COP24 in Katowice, Poland, in 2018. It awaited APA to complete its work to become a functioning conference for a to-be-implemented agreement.

In place of the KP, a twenty-five-page treaty with instructions on the annual carbon emission reductions expected of the most advanced economies, we now had the Paris Rulebook, well over one hundred pages, with many parts as yet not resolved, which was to serve to ignite individual, collective, public, and private climate action across all Parties. Reduction and mitigation of carbon emissions was only a part of it. I could see why Hafij would be drawn to the KP over the PA. The burden of responsibility and the pathway to mitigating climate change were clear within KP. Many of the activists with whom I spoke were mournful of the KP's fate, indicated by their insistent calls to the developed Parties that had ratified it to meet their pre-2020 commitments. The phrase "pre-2020" was code for signaling that those efforts at reduction as mandated under the KP ought not to stop while Parties geared up for a new round of commitments and actions to take effect after 2020 with the implementation of the Paris Agreement (Rogelj et al. 2015; Averchenkova and Zenghelis 2018).

Not everyone felt the same as Hafij. Another acquaintance, Mark Jariabka, a US-based lawyer whose organization worked on behalf of the AOSIS, said good riddance to the KP. According to him, it had hindered any meaningful climate action through its exclusive focus on counting gross carbon emissions and even that only by a subset of countries. With every Party to the Convention signed on to the PA, one could finally imagine the scale of integrated action needed to combat climate change. With everyone doing something, the needle on the temperature gauge could be expected to move down. But, he added in his usual dry tone, the agreement was too little, too late. Had the agreement come at the start of this process, in the early 1990s, then one could have imagined realistic change. But coming as it did in 2015, as the world was increasingly locked into a 2+°C change in temperature rise from the pre-industrial period, the PA would not be able to reverse the inevitable, at least not in the foreseeable future.

I found both gestures, the mournful last glance at the KP and the affirmative yet skeptical appraisal of the PA, compelling. I had heard criticisms of both legal instruments before. I also knew that both lawyers were practiced pragmatists, as all those involved in the negotiations were. They were going to work with whatever evolved regardless of their individual feelings of ambivalence. What was communicated in their comments were different approaches to climate change, one premised on the elegant economy of actors and actions and the other on a more participatory approach to action. The first emphasized a top-down compliance regime and the second a voluntary growth of international norms from the ground up.

Yet the two treaties were not that different in organization. Both relied upon reporting and review to connect national actions to the global arena. They subscribed to (1) reporting, that is, the means by which Parties communicated their greenhouse gas (GHG) inventories and biennial reports and four yearly national communications on what they were doing to mitigate, and (2) review, in which the submissions were reviewed through a desk study by the Secretariat and an in-country visit by technical experts, followed by a multilateral assessment in which each Party's reports were submitted to Q&A by other Parties under the auspices of the Subsidiary Body for Implementation, one of the two permanent subsidiary bodies that supported the Secretariat's work. But whereas the KP's reporting and review was required for Annex 1 countries, only recommended for some non-Annex developing countries, and not required for the LDCs and SIDS, with the PA reporting and review was required for developed and developing countries and recommended for LDCs and SIDS (Yamin 1998; Dessai and Schipper 2003; Bodansky 2016; Falkner 2016; Klein et al. 2017; Kemp 2018).

To understand the almost unanimous hope pinned on the PA, one must attempt to distinguish the theory of change embedded in the PA in comparison to the KP. I use the phrase "theory of change" because my interlocutors often used it.[1] It meant "projected change" and the policies needed to be adopted in the present to bring about such change. It was different from saying "goals and outcomes" in so far as "theory of change" also took into consideration that attempted change could lead to opposite reactions and contingent effects. In other words, this orientation to change was self-aware, self-assessing, and, potentially, self-correcting. To draw out the theory of change embedded in the PA in comparison to KP was to draw out the widest field within which it sought to operate, which might be called its ideational side, and the ways it attempted to circumscribe its scope to be operational, which might be called its practical side. Here I draw on anthropological modes of reading texts, which pay attention to the formal and historical features of a genre and how a text embeds aspects of the situation or context in which it was created as well as retains an independence from it (Barber 2007).

Culling insights from lawyers, in this chapter I compare the discursive thrust and motivations for the Preamble and articles of the Paris Agreement (PA) with the Kyoto Protocol (KP), which the PA effectively replaced. The triad of accounting, accountability, and transparency at the heart of the PA, by which each Party's contribution to climate action was to be monitored after the agreement goes into effect, indicates that it might not be the most ambitious response to climate change. However, good bookkeeping was understood as necessary for building trust among Parties of the Global South and North within the current geopolitical order before climate could become a priority for everyone.

First, a word about international law. It is important to keep front and center the fact that international law is only as strong as the will to uphold it (see Hernandez 2012; d'Aspremont 2012). Unlike national law, in which an already constituted legislative body creates and enacts the law, which then acquires an autonomy of its own, in an international legal instrument, the law and the Parties to it are co-constitutive. Thus, the opening words of the Preamble of the PA, "The Parties to this Agreement," called up both the agreement and the Parties, and they existed only in relation to each other. There was no agreement without the Parties, and there were no Parties to the agreement without the agreement.

It follows from this that the PA could only act on Parties in the way the Parties allowed. So, for example, when we read in its text that the Parties "shall" do something, it effectively meant that the Parties had given the PA the authority to hold them to an imperative. When we read that the Parties "should" do something, in that case the PA had been authorized to enjoin an action without mandating it. This prevailing situation made international law only ever normative rather than prescriptive. Even when there was language of the imperative within it, Parties could modulate international law's impact on them through a whole range of responses, from merely taking the imperative under advisement, to ignoring it, or even withdrawing their consent to be shaped by the law. However, such action was considered extreme by the historical norms that had evolved among the countries that made up the global political order (see Tomz 2012). That's why the US decision to first withdraw from the KP under George W. Bush and from the PA under Donald Trump shocked the international community. Despite the United States' fitful commitment to global climate policy, withdrawal indicated a reneging of rules of conduct, making the United States into a rogue nation whose behavior could no longer be predicted, much less modulated through appeals to compromise and peaceful planetary coexistence (Zhang et al. 2017).

The Principled Side of the PA: The Preamble

In his book *The Great Derangement* (2018), the novelist and cultural critic Amitav Ghosh indicts literature for its failure to communicate the experience and imperatives of climate change. Interestingly, he considers the Paris Agreement within the scope of his understanding of literature and provides a reading of it. Referring to the PA's lengthy Preamble, which *recalls, further recalls, welcomes, recognizes, decides,* and *requests,* he faults the PA for "confinement and occlusion . . . concealment and withdrawal" (154–55). He finds the agreement to be tepid in naming the condition to which it is the solution (see d'Aspremont 2012). He also finds the imagination of change, that is, market-based solutions with reliance upon the very same private sector that created and intensified the problem of climate change in the first place, as problematic in both economic and moral terms.

While I agree with Ghosh's general critique of the PA—and activists within the process have long said similar things—I find his reading of the text of the agreement attributes too much coherence to it. Rather, guided by Mathew Stilwell, an Australian lawyer who was part of the process in many ways as activist, advisor, legal advocate, and teacher, I read the text as heteroglossic—often speaking in multiple voices and expressing different viewpoints (Bakhtin 2013).[2] Beginning with the Preamble, Mathew walked me through elements of the agreement, pointing out the voices and viewpoints embedded within it.[3]

The Preamble consists of the sixteen elements that come after the opening words, "The Parties to this Agreement." It is largely symbolic; that is, it articulates the principles good to keep in mind when putting the PA into effect but not required to be implemented (see Piotrowski 2011). The verbs employed in the Preamble, such as "recognizing," "taking full account," "emphasizing," "acknowledging," "noting," and "affirming," suggest a passive receiving rather than active undertaking of any action.

The first part of the Preamble, "*Being* Parties to the United Nations Framework Convention on Climate Change, hereinafter referred to as 'the Convention,'" puts the Paris Agreement squarely within the provenance of the Convention. The wording is important to emphasize because the buzz had been that the PA was crafted not only to take over from the KP but to displace the importance of the Convention for the process. Mathew drew my attention to how while decisions on the KP took their directive from the Convention, those on the PA only ever gestured to the Convention as a source of inspiration. Later, Bill Hare of Climate Analytics would dismiss such concerns as conspiratorial thinking indicating an overzealous fidelity to the Convention.

The Convention is once again acknowledged in the third part of the Preamble with the reference to the "principle of equity and common but differentiated responsibilities and respective capabilities." To remind the reader, the principle of CBDR, perhaps one of the most capacious but also the most contested that I have introduced earlier and have been discussing throughout this book, refers to an understanding arrived at the 1992 Rio meeting where the Convention was finalized. The shared understanding was that developed countries accepted responsibility for producing anthropogenic climate change

and agreed to take leadership and the lion's share of the responsibility to mitigate its effects, but with the understanding that every country would eventually take up more responsibilities, albeit differentiated according to national
circumstances. The bone of contention that had arisen since was whether
the emphasis ought to lie on the first half of this shared understanding, that
is, that developed countries had to mitigate *immediately*, respective of what
anyone else did, or on both together, that is, that developed countries would
mitigate, and more than others, *only after* responsibilities had been apportioned
among all countries. This contestation (some would say obfuscation) of the
original treaty endured in the present. In my interviews with Richard Kinley,
the retired deputy director of the UNFCCC (introduced earlier), he expressed
grave disappointment that developed countries hadn't taken leadership, as had
been hoped of them, in the early years of the process. Had they done so, the
world would be at a different place now in terms of the climate crisis.

The fourth part of the Preamble further implied the Convention by urging
that action was needed based on "the best available scientific knowledge." Science was the call to action that had prompted the original 1992 Rio meeting.
And it was the incentive for climate action largely accepted by all Parties, until
the post-truth era ushered in by Trump, when attempts were made to down-
grade science to opinion.[4]

Aside from these few nonspecific mentions of the Convention and one reference in the second preamble to the Durban Platform for Enhanced Action,
the working group that drafted the text of the agreement, the Paris Agreement
did not mention any other legal instruments or decisions in justifying itself,
either as its inspiration or even simply its precursor. In contrast, the KP recalled
specific provisions and articles of the Convention, in addition to the Berlin
Mandate that instigated it. It was as if the PA was in a society largely unto itself,
with a vague historic connection to the Convention and no mention of the KP
that preceded it.

Parts 5 to 7 of the Preamble spell out the national realities making climate
action both urgent and difficult. They express acknowledgments by the Global
North of the challenges faced by the Global South and indicate differences
internal to the Global South. Mathew said that the Global North, or at least
the EU, wanted to play the part of the good guys in acknowledging the special challenges of countries in the Global South but also wanted to indicate
that there was a limit to how much help they could proffer.

The problem with framing the issue this way was that it made attempts to
fight climate change an individual affair aided by others further along in their
efforts. Mathew didn't speak of the historical responsibility of the Global North
for creating the problem in the first place or its climate debt to the rest of the

world. Rather he just spoke of the fact that this mode of individualizing responsibility and buffering it with humanitarian aid erased the wider system in which everyone was operating. Later I realized I should have asked him what he meant by the wider system. Did he mean the free-market economy, or the wider-reaching concept of capitalism, or some other political system incorporating them, such as capitalist democracies? Searching the web, I came across an interview Mathew gave to his hometown paper in July 2016, where he said, "It's not just a problem of emissions . . . it's a problem about a system that promotes particular lifestyles, consumption patterns and injustices" (Cica 2016).

With these comments in mind, I read the fifth and sixth part of the Preamble as parsing out the countries in the Global South to ascertain who was most deserving and how. The first recognizes the specificities of "developing country Parties." Given that the market economy was the pinnacle of development within the current global economic system, the needs of developing countries in their effort to transition into it was hereby noted. At the same time, given that the global economy was not based on cooperation but on competition, there was also a hint of anxiety that the Parties being sought to be protected within the scope of the Preamble were potential future competitors, if not already so in 2015 (as opposed to the time of the Convention in 1992). Thus, the special accommodation was not a blanket one, being qualified by the phrase "especially those that are particularly vulnerable to the adverse effects of climate change."

The sixth part of the Preamble of the Paris Agreement further parses out the countries in the Global South. The 1992 Convention differentiated developing from developed Parties. Although the least developed countries category is a longstanding one within the wider UN, there was no mention of least developed countries within the Preamble of the Convention. The same held true for the KP. The PA maintains the distinction between developed and developing, but it also mobilizes the category of least developed countries. And, unlike the KP, which mandated that the most advanced developed countries help all developing countries, it hints that help in the form of finance and technology would be restricted to least developed countries: "*Taking full account of* the specific needs and special situations of the least developed countries with regard to funding and transfer of technology."

The PA reverses the previous order of priorities. For the KP, the economy is the basis for divvying up the countries of the world to ascertain who had to do what with respect to climate mitigation. The endgame was climate mitigation. For the PA, one mitigates to allow economies to progress. It was mitigation not for climate's sake but to sustain or grow the global economy. The PA, thus, appears to prioritize the market economy over creaturely existence (see Ciplet and Robbins 2017).

Steven Bernstein's *The Compromise of Liberal Environmentalism* (2001) provides some background on this shift in priorities between the KP and the PA. In the early phases of the UN's efforts to launch a global response to environmental pollution and degradation, there were several notable points of contention. The first was between those developed countries that wanted a command-and-control approach to environmental regulation, by which economies were to be brought into line through bans, tariffs, and targets, and those that favored a market-friendly approach, whereby markets were incentivized to adopt environmentally friendly approaches and products.

The second point of contention was between the Global North, which presumed environmental protection was a common good for all, and the Global South, specifically developing countries, which bristled at this presumption. The latter pointed out that the North was responsible for their states of underdevelopment. Only after overcoming this would they be in a position comparable to the North to protect the environment. And rectifying the situation required the right to development, access to resources, protected access to the market, as well as recognition of their national sovereignty and rights to exploit their own resources. In effect, developing countries demanded the right to pollute by laying claims to what remained of the global commons (the atmosphere, oceans, etc.), while expecting those countries who had already benefited from it and continued to take from the commons, despite having used up their fair share, to be penalized or taxed. This position will sound familiar to readers, as it was the one espoused by Lidy Nacpil, one of the activists highlighted earlier.

Bernstein claims that the second position was considered untenable within the global political order, with the orientation toward a market economy supported by the neoliberal ethos of individualized responsibility. Developed countries closed ranks by having the regulation-driven group capitulate to the market-friendly group. As a result, a market-friendly approach to climate change came to dominate the UN process. Even developing countries came to accept the superiority of the market-based, voluntaristic mode of fighting climate change, despite that fact that it was none too favorable for them, given the structural inequalities built into the global market. Rather they were brought around to the market-friendly approach with promises of treating their sovereignties as sacrosanct, acceding to them the right to exploit their own resources as they wished, the right to development, and some limited protections within the market, such as exemptions from tariffs, price stability, etc., but with no serious redistribution of the captured rights to the global commons.

Bernstein further shows how "sustainable development" emerged as the compromise solution. It encouraged taking the environment into consideration by making it into a market opportunity within existing economic practices, for instance, through incentivizing pollution control in the North. It made poverty alleviation, a central tenet of development in the South, the equivalent of controlling pollution in the North. A developing country effectively undertook sustainable development by doing poverty alleviation. And it meant developed countries could offshore the restructuring of their own economies through investment in the economies of developing countries. Bernstein shows how this implicit settlement was reached as early as 1987, within the important UN-sponsored Brundtland Report that initiated the Rio Conference (see also Jacob 1994).

Given this backstory, it should not be a surprise that parts 8 through 10 in the PA's Preamble emphasize "sustainable development," "eradication of poverty," "safeguarding food security," and "the creation of decent work and quality jobs," with nationally defined development priorities in mind. Yet, despite the PA's acknowledgment of the right to development of developing countries and, presumably, least developed countries, those of the Global South often professed preference for the Convention over the PA. While there were many reasons for this (such as the Convention's strong support for equity and CBDR), one that is readily apparent through reading the texts of the two together is the vast difference in representation of developing countries across them. In the Convention, developing countries were presented as being like other countries—seeking resources and energy efficiency to achieve sustainable development, just behind the curve and in need of a leg up. In the PA, developing countries, represented by a much smaller subset of them, are quite distinctly set apart from the rest of the world. Presented as plagued by poverty, hunger, vulnerable food systems, and a redundant workforce, they appear almost pathological, in need, and, indeed, deserving of help but simultaneously, one suspects, not likely to be able to leverage such help. Were these, then, potentially sacrificial zones (Nel 2015), that is, those parts of the world that would have to be considered lost causes and abandoned in the future? I began to appreciate the words said to me in passing by Seyni Nafo, the Malian head of the African Group of Negotiators, when I asked him what the Convention meant to him. He said, "The Convention is God."

While the majority of the Preamble considered so far indexes Party interrelations, the remainder demonstrates the influence of civil society, structured into constituencies, including environmental groups, trade unions, women, the young, business, and the indigenous, upon the negotiation process. They are more moral in nature and proliferated figures of pathos to inspire action

(Riles 1999a, 2001). Preamble part 10 speaks for trade unions asking for the just transition of the workforce as economies changed to mitigate climate change. Part 11 shows coalition building across several constituencies, with an all-encompassing statement urging Parties to take into consideration "human rights, the right to health, the rights of indigenous peoples, local communities, migrants, children, persons with disabilities and people in vulnerable situations, and the right to development, as well as gender equality, empowerment of women and intergenerational equity." And 13 throws in concern for ecosystems, oceans, and biodiversity, invoking Mother Earth and climate justice (see Borie and Hulme 2015). The latter is in quotes and prefaced with: "noting the importance for some of the concept of 'climate justice.'" The qualification refers to the lack of consensus of what this concept meant but also the strength of the push to include it regardless.

It's quite an assortment of asks, and it shows concerted efforts by constituencies to represent the interests of a wide swath of politically marginalized groups. In conversation with Sebastien Dyuck, a senior attorney for the Center for International Environmental Law (introduced earlier), he recounted to me the years of organizing before COP21 that went into bringing all the constituencies together and putting pressure on negotiators to include these joint statements within the Paris Agreement. He pointed out that while members of these constituencies, many of whom hailed from the Global South, wanted to uphold the needs and rights of the South, they also sought to include protection from their own governments within these statements. Here we sense a tension between developing countries' demand for acknowledgment of their sovereignty and the fear that such a blanket acknowledgment would give them impunity to do as they wished within their territory, against their own populations. A possible critique of developing countries by developed countries found a weak link to the actual experiences of the poor and the marginalized in developing countries without good human rights track records. And it makes this set of Preamble statements the most interesting to follow over the course of negotiations, because it gets people out of their usual categorizations as, say, official delegates, environmental activists, or legal experts, to experience and express contradictions within their Party positions, ethical commitments, and acts of political solidarity. Dyuck's account of how human rights fared within this process, sketched earlier, provides an important vantage on the passage of these principles from the agreement to the implementation guidelines, or "Rulebook."

Parts 14 and 15 are also oriented toward civil society. The former emphasizes the need for education on the PA, which is a frank admission of the complexity of the issues covered by it. And the latter recognizes the importance of engagement by "all levels of government and various actors" in addressing climate change while acknowledging that there may be limits upon such wide participation by the national legislations of Parties. The final part, the sixteenth, states that sustainable lifestyles within developed countries could also help address climate change. This last is interesting because it pinpoints the wasteful production and consumption that underwrites high standards of living in developed countries and puts the onus on their citizenry to pursue more sustainable patterns. But it also individualizes responsibility, offsetting the responsibility of states. Furthermore, sustainability is understood by a very limited set of actions, specifically lifestyle and consumption changes. Such actions would largely work to incentivize the market economy to be environmentally friendly. In other words, the end goal of the Preamble is still to bolster the market economy. This isn't quite the system change Mathew was calling for.

The twelfth part of the Preamble stands out because it introduces a consideration of nature, albeit only in the forms of sinks and reservoirs of carbon: "*Recognizing* the importance of the conservation and enhancement, as appropriate, of sinks and reservoirs of the greenhouse gases referred to in the Convention." Here the PA recognizes the need for sinks that absorb greenhouse gases and the need to keep intact reservoirs or sources that were currently storehouses of greenhouse gases. The PA is thereby harnessed to the KP because while the PA cited the Convention as its reference for this directive, it was the KP that had operationalized it.

After the part about sustainable lifestyles, the Preamble segues into the main text of the PA. Although one saw much jockeying by various human and nonhuman constituencies for inclusion, suggesting how the lively landscape of negotiation took up home within the text, it is also clear from it that we were moving out of the world forged by the Convention and KP into a new one to be created by the PA. Its newness is signaled by the reduction of the scope of CBDR and the PA's inclusion of developing countries among those having to bear responsibilities for mitigation beyond the goals of sustainable development and poverty alleviation. And, as Ghosh had sensed from his reading of an early draft of the PA and I found to be the case from my reading of the Preamble, it was clearly market oriented. As the US secretary of state John Kerry indicated in his speeches, it was a signal to the market to indicate that alternative green energy was here to stay.

It remains unclear from the Preamble why developing countries would consent to this curtailing of their hitherto protected status within the negotiations and, if they found hope in the PA's growth paradigm, how they saw themselves benefiting from it. It may be useful to recall a bit of negotiation history to understand this paradox. While COP15 in Copenhagen in 2009 was widely considered a failure for reasons that have already been explored in earlier chapters, it was important in so far as it included developing countries among those to be involved in future mitigation efforts. And there was a promise of "fast-track finance" of $30 billion by developed countries to be distributed among developing countries by 2012 to get them in shape to take on mitigation and adaptation goals, with a further promise of at least $100 billion to be raised per year from 2020 onward toward the anticipated increasing costs of climate change's effects (Schalatek, Bird, and Brown 2010; Stadelmann, Roberts, and Huq 2010).

The larger amount didn't materialize in 2020, with developed countries backpedaling by saying that they had hoped for a stronger market response to make up the difference between outright grants and loans and the $100 billion benchmark. However, this bit of history makes it likely it was the promise of finance that brought developing countries to the table. My reading of the main text of the PA, guided by the comments and insights of interlocutors, suggests that there was sufficient ambiguity within the text to allow for deferral, if not outright reneging by developed countries on their promises of finance (see Roberts and Weikmans 2017). So there had to be something else that was new to the PA, something more at stake within it for so many countries to agree to it.

The Practical Side of the PA: "The Deal Is a Strange One"

While I went through the Preamble with Mathew and later with Sebastien and others, discerning the presence of many within it, the actual text of the agreement was much harder to read and decipher. Although Ghosh decries the PA's tendency to conceal or occlude the worst of climate change, I find it to be more circumlocutory, almost like it was written as spoken, and purposely so. It is chatty in tone, with many apparent digressions, loops, and repetitions. In some instances, there are long strings of words to specify a task, making the task itself seem interminable. In other instances, there are strings of qualifications, making the task seem very specific. It reads as skittish and nervous, as if something might be forgotten, repeating from article to article, sometimes even from section to section within an article.

One reason given for the meandering quality was that the text came after many years of strained relations among Parties after the failed negotiations in Copenhagen. Consequently, the French foreign minister Laurent Fabius, in charge of delivering an agreement in Paris in 2015, sought to be as inclusive as possible. Negotiators kept the text deliberately vague, making it what is called "constructively ambiguous," so that everyone could notionally agree (Geden

2016). That meant they deferred the inevitable task of giving more definition to the work that had to be undertaken to a later date, to the Rulebook of the Paris Agreement that had to be negotiated so that the agreement could be implemented by 2020, five years after it was signed into being.

The PA boasts twenty-nine articles, the most relevant ones being 4 through 15. In the first few years of attending the COP, I struggled to memorize the articles so that I would know what Article 4.1 or 6.4 referred to. Invariably, I would find myself opening the document to seek out the relevant article cited in whatever conversation I was listening in on. Over time, I came to know that Article 2 articulated the principal aims of the agreement, Article 3 brought up nationally determined contributions (NDCs) as the main way by which countries were to contribute to climate action, Article 4 referred to mitigation, Article 5 to the maintenance of sinks and reservoirs of greenhouse gases, Article 6 to cooperative approaches by which to meet one's voluntary contributions (which was code for carbon markets), Article 7 to adaptation, Article 8 to loss and damage, Article 9 to finance to be provided by developed countries to developing ones, Article 10 to technology to be transferred from developed countries to developing ones, Article 11 to the need for capacity building within developing countries, Article 12 to public education on climate change, Article 13 to a transparency framework by which developed and developing countries were to maintain the standards of reporting, Article 14 to the global stocktaking on progress toward meeting the goals of the agreement, and Article 15 to facilitate implementation and compliance with the agreement.

Even as I prided myself in knowing the articles, I was thrown into a panic when I realized that in addition to referring to the agreement, people were also referring to the COP21 decision text (Decision 1/CP21), which provided further instructions on how to prepare the agreement for implementation. Also, the APA tasked with producing these implementation guidelines reorganized the articles by agenda items of its own numbering. I cried out in frustration when I finally comprehended that at any given time people could be referring to an issue, such as mitigation, as articulated in Article 4 of the Paris Agreement, or as discussed in CP21 numbers 22–40, or as elucidated in APA agenda item 3.

I read whatever analyses I could in the endless reams of coverage to be found in newspapers, journals, organization reports, and blog posts, making sure to be attentive to who was saying what to get a sense of the myriad interests directing the analysis. For instance, in the case of media coverage, I quickly realized that while the *Guardian* was excellent in terms of providing a diversity of perspectives, one read the *Economist* if one wanted clarity on the economic issues involved and *Forbes* to get the business sector's read of the negotiations.

The *Atlantic* had a deconstructive approach to the Paris Agreement, and I found myself returning frequently to Robinson Meyer's article "A Reader's Guide to the Paris Agreement: The Most Important Piece of International Diplomacy in Years, Deciphered" (Meyer 2015b).

Meyer, a staff writer for the *Atlantic* and author of the *Weekly Planet*, a weekly newsletter, declared the deal (PA) a strange one, most immediately because of the clear disjunction between how it is portrayed by US leaders like Kerry, for whom it was merely a spur for private investment rather than a legally binding document, and the august standing it had internationally, in keeping with the amount of diplomatic finessing that went into ensuring that the global community of nations would agree to be Parties to it. He shows how the text is oriented toward providing complicated and interlinked responses to important questions concerning climate change. According to Meyer, these questions are: By how much should the world limit warming? How quickly will the world abandon fossil fuels? Who should pay for the costs of climate change, and how much should they give? How often should nations check and reassess their emission reductions? Who should make sure nations meet their reduction goals? Who is responsible for the loss and damage caused by climate change? Who bears responsibility for protecting the climate?

The answers to these questions are within the PA. For instance, he says that to avoid an impasse on deciding when global greenhouse gas emissions should peak and start going down, the PA opts for the safe "as soon as possible." He immediately picks up on the fact that climate finance is never coming by the fact that the PA and COP decision text use the language of "mobilize," which is no guarantee that it will happen. He shows how "stocktaking" and "ratcheting" are two different means by which countries would be expected to gauge if they are doing enough, with the second providing a means to continually increase one's ambitions within one's reported nationally determined contributions. He shows how instead of an external organization ensuring compliance, the process and the Parties within it would be subject to a common framework on transparency. Upholding transparency requirements will ensure compliance at the ground level. Meyer is also attentive to the dynamics between developed and developing countries, particularly around the issues of CBDR, finance, and loss and damage.

While cognizant that the agreement is nowhere near as ambitious nor its status as secure as it ought to be for one that was to deliver humanity from destruction, Meyer does think that the PA introduces newness into the climate process through "a new cycle, a new calendar of cutbacks, onto the future," which will "help decide . . . how the human species addresses a crisis on its home-world in at least the next quarter century" (Meyer 2015a). Although

I agree with Meyer that there is more going on within the PA than Ghosh gives credit, he underestimates the nature of newness within it. The interlinkage between accounting, accountability, and transparency shows how norm setting the PA aims to be. Here I consider Articles 3 and 4 in greater depth to illustrate my insights into how the PA achieves this.

Article 3 of the PA indicates that Parties "are" to undertake and communicate their most ambitious efforts as their contribution in the form of "nationally determined contributions" (NDCs). While within treaty language, "shall" indicates an imperative and "should" a strong recommendation, "are" and "will" indicate a class of actions on which there is collective agreement. NDCs were the newest form of Party communication within the UNFCC. In addition to the greenhouse gas inventories, biennial reports, and four yearly national communications, all already part of the ongoing reporting regime, Parties to the agreement were asked to provide five yearly reports on what emission reductions they voluntarily intended to undertake and the pathways to do so. What made them different from earlier submissions was that the NDCs were Party-led submissions of emission reduction goals, as opposed to externally imposed goals, as in the KP. They were future oriented, projecting what the Party would do, rather than an accounting of what had been done. Parties were encouraged to "ratchet" up their ambition with each subsequent NDC, their progress measured relative to their previous submissions. Furthermore, review would no longer be a multi-Party, Q&A assessment but rather a facilitative sharing of views to move from critical evaluation and judgment toward mutual encouragement.

Article 4 on the mitigation provides a clear example of how NDCs were to be operationalized. This article alone had nineteen numbered sections. While section 4.1 reminded that to reach the 2°C goal countries must have their emissions peak as soon as possible, with nods to developing countries needing more time, section 4.2 went straight to the communication of NDCs: "Each Party shall prepare, communicate and maintain successive nationally determined contributions that it intends to achieve." While the phrase "intends to achieve" would suggest that the Party should both report and act on its NDCs, the triple emphasis on reporting, through "prepare," "communicate," and "maintain," made it clear that Parties maintaining good records on their pledged contributions was as important as maintaining the level of action necessary to meet their voluntary targets. This ambiguous wording produced the sense that climate action was both mitigation and reporting on mitigation. Actual action was paralleled or even overshadowed by its enumerated representation within the PA, making representation more important than the actual world to which it referred.

A brief look at the history of accounts keeping and accountability is useful to understand the importance given to reporting and review by the climate negotiation process. Jacob Soll in *The Reckoning: Financial Accountability and the Rise and Fall of Nations* (2014) writes that the relationship between accounts and accountability evolved over a long period of time. Double-entry bookkeeping emerged with the rise of mercantilism and the need to gauge profit and loss. It took some time for mercantile practices to be taken up by heads of government, as rulers feared that records keeping of expenditure could undermine their authority. Thus, even when national treasuries put in place meticulous bookkeeping, it was shielded from public view to protect the ruler from public scrutiny. Furthermore, secrecy enhanced the king's power. It was only over time that the understanding emerged that accounts could be used to show one to be accountable to a wider public, thereby strengthening a leader's standing and support. As a result, accounts keeping went from being a commercial practice and practice of statecraft to being the sign and standard of responsible government. The climate negotiation process drew upon this common connection between accounts keeping and accountability in its expectation that the Parties will regularly report on their activities and, thereby, open their activities to review (Chayes and Chayes 1998; see also Hoffman 2016; Park and Kramarz 2019).

Section 4.5 assured developing countries that support would be forthcoming to help them raise their ambitions regarding targets. Again, this subsection was ambiguously worded. It was unclear whether the support was to help developing countries be more ambitious in their actions or help them fulfill the reporting requirements as laid out for NDCs: "Support shall be provided to developing country Parties for the implementation of this Article, in accordance with Articles 9, 10 and 11, recognizing that enhanced support for developing country Parties will allow for higher ambition in their actions."[5] Section 4.6 gave special dispensation to least developing countries and small island states, giving them the option to prepare and communicate "strategies, plans and actions for low greenhouse gas emissions development reflecting their special circumstances" if they so wished.

Section 4.7 allowed adaptation actions that yielded mitigation co-benefits, that is, emission reduction, to be counted within country-specific disclosure of efforts to mitigate. Thinking of adaptation as potentially contributing to mitigation goals brought the two pillars of climate action together. It raised interesting discussion within negotiations over the Paris Rulebook as to whether mitigation and adaptation should be reported together within the same tabular format. Would reporting adaptation efforts within the scope of mitigation compromise mitigation, by allowing Parties to have a soft, offshore option of supporting development projects or utility installations elsewhere over under-taking economy-wide emissions reductions at home? This kind of sharing of emission-reduction burdens had been previously undertaken through the Clean Development Mechanism and Joint Implementation within the KP but was faulted for many reasons, ranging from double counting carbon credits, once by the country undertaking the project and a second time by the country that underwrote the project, to questionable projects that did not deliver on any meaningful climate action but just met the letter of the law (see Murray 2000; Cooper 2001; Schneider 2011).

One could argue that the thrust of this subarticle is less on how adaptation could benefit mitigation and vice versa and more on ensuring that bringing the two together didn't encourage creative accounting (see Van Asselt et al. 2016). Another fear was whether shared reporting would lead emission-reduction goals to displace the developmental ends of adaptation. After all, what would be the shared metrics between the actions? Emissions reductions were measured in terms of metric tons, global warming potential, and radio-active forcing values, as in the standards for carbon inventories provided by the IPCC, but adaptation presumably was measured with an entirely differ-ent set of metrics, like studying different kinds of vulnerability and potential for resilience. Would every development project now have to be recast in the terms of emission reduction (Lippert 2013)?[5]

The third fear, the one most often articulated by developing Parties in ne-gotiations over the Paris Rulebook, was that not entwining the two in terms of finance prevented a comprehensive understanding of what developed Parties were doing to combat climate change (Van Asselt et al. 2016; Weikmans and Roberts 2019). Developing countries wanted developed countries to be explicit about what they had achieved and aimed to achieve in terms of mitigating car-bon emissions, indicated by the phrase "transparency of action," but did not feel that they were obligated to do the same. And they wanted the North to be clear in terms of what new and additional funds they had directed to abating climate change or adapting to it without either double counting development funds or redirecting such funds to climate-oriented projects, indexed as "trans-parency of support." Developed countries also sought transparency of action but wanted developing countries to take on the same reporting obligations for mitigation to allow for comparison across developed- and developing-country actions. They were less keen on the transparency of support on the grounds that it imposed more of a reporting burden than was needed. The most vul-nerable and least developing countries were happy to be let off the hook from taking on too many reporting burdens.

What these deliberations revealed was a deep-seated distrust on the part of the Global North that the Parties of the Global South couldn't be trusted to be accountable, if they were indeed deserving of aid at all, and, on the part of the Global South, that the Parties of the Global North were possibly double-, if not triple-, counting what they gave in development aid (as adaptation assistance and carbon credits procured to count toward their national GHG inventories). Wariness on both sides threw the purported link between accounts keeping and accountability into question. If previously it was one's accounts, visible for all to see, that kept one accountable, without trust in one another the same accounts were awash in suspicion. Even though they were visible, they were suspect for not being legible, for hiding a lack of commitment, inaction, and even fraud in plain view.

Section 4.8 of the PA appeared to address this deficit of trust. It required that the Parties practice "clarity, transparency and understanding" in all their reporting, particularly in their NDCs. In effect, the PA introduced the language of transparency into the discursive universe of the Convention and the KP, almost making a fetish of it by including it in every article and giving it its own article (Article 13; see Ciplet et al. 2018). Historically, as Mary Poovey writes in *The History of the Modern Fact* (1998), transparency was considered an intrinsic aspect of accounts keeping because of the neutrality imputed to numbers over spoken and written language. However, as numbers too came to be suspect, transparency acquired a value somewhat independent of numbers. It came to mean making a morality of being open. One didn't just keep visible and legible accounts to meet an external standard for being accountable. One did it for its own sake. The Paris Agreement mobilized transparency to address the problem that there was no trust, neither in numbers nor between Parties. It allowed for the acknowledgment that there were structural problems, deep-rooted inequalities, and unfairness within the existing setup and that there was likely good reason for mutual distrust. But it also held out the promise that if everyone espoused transparency as a virtue, then there was more likelihood of accountability to one another by means of good bookkeeping.

PA and KP: Comparing Apples and Oranges?

Throughout this chapter, I have been exploring PA's convergences and divergences from the KP. In my reading of the Preamble, what struck me was that the PA appeared to be more market oriented than the Convention and the KP. However, Bernstein's *The Compromise of Liberal Environmentalism* (2001) reminds us that the Convention and the KP were crafted during the period when the market had already come to establish itself as the way to fight climate change. In other words, it wasn't as if the KP didn't rely upon the market for mitigating climate change but that the PA was just more neoliberal in its orientation, with its push to disaggregate historical groupings (such as developing countries into smaller deserving groups) and to individualize responsibility.

The PA, too, continued the reporting and review procedures initiated by the Convention and the KP, with some new additions, such as NDCs and the periodic global stocktaking and the transparency framework, and tweaks, such as the move to a facilitative rather than a punitive compliance regime. I would argue that the inclusion of transparency should clue us in to the fact that the PA was operating in a very different global context than the Convention and the KP. It was one that was marked by greater distrust and the failure of reports to provide accountability to one another. The inclusion of transparency in every dimension of the PA indicated this situation and the PA's attempt to counter it.

The promise of the PA was to help all Parties adopt the same accounting regime. This is indicated by the circumlocutory way the PA states its objective, which is to "enhance" implementation of the Convention and "to strengthen the global response." And it's there in every one of the articles in which there is ambiguity as to whether the PA is going to help Parties undertake actions or help them to account for their actions. It is there in the way that accounting is central to mitigation NDCs, carbon inventories, and market approaches and to adaptation NDCs and loss and damage.

What, then, is the theory of change in the PA? If one understands the KP as the means by which specific Parties were mandated to take steps to combat climate change, while others were recommended or excused, the PA is about making sure that the Parties have a modicum of trust in one another. It is based on the presumption that only a shared accounting regime informed by transparency can produce this trust and that without this trust, there will be no effective action to combat climate change. The PA is, in the end, an agreement to account together for the sake of the planet.

6

A Thrice-Told Tale of Negotiations

There was a lot of buildup to COP24 in Katowice, Poland, in 2018. After the PA had been ratified in Paris, negotiating the rules for its implementation started at the intersession in Bonn in 2016 and had carried on through an additional intersession in Bangkok added to the calendar and scheduled for just a few months before COP24 to hasten progress. The rules of implementation, previously called the Paris Rulebook, had to be adopted at COP24, according to the terms of the PA. One thing I had learned by observing this process was that deadlines were met even if the ensuing actions or products were hollow. In Katowice, witnessing the Polish government's very visible suppression of civil society organizations, Nathan Thanki, the organizer of DCJ and a veteran of the process, summed it up as an indication of the Polish COP Presidency's determination to deliver on the rules. He even speculated that the suppressions were spectacles to throw activists off the scent of the heavy-handed measures the Polish Presidency intended to take within the negotiation process to bring about a decisive outcome at the meeting, which was being billed as "Paris 2.0." For Poland to deliver the rules would be no small measure of the successful leadership of the COP and demonstrate that a Party could achieve results without too much diplomacy, as vaunted by the French.

Developing countries were increasingly worried that the Paris Agreement was created to set aside the KP and even perhaps the Convention. Pre-2020 conversations were not progressing, and it did not look like enough countries would ratify the KP for a second commitment period. They fretted that CBDR, the one principle within the Convention where questions of historical responsibility, climate debt, justice and equity issues, and rights to development were addressed, however inadequately, was going to be watered down. They felt that although the PA had come with a treaty text and a substantial decision text to accompany it, everything was being negotiated again at the follow-up sessions. The push to make every bit of the PA clearer made sense to them with respect to the transparency of action and support provided by developed countries, but the same push applied to developing countries was seen as unfair, burdensome, and possibly infringing on their sovereignties. They feared that more of them would have to take on mitigation targets, which would have consequences for their national economies and efforts to achieve decent standards of living for their populations. And then there were the "blockers," "laggards," and "obstructers" among them who had been holding up progress for decades (Depledge 2006, 2008; Kemp 2016; Urpelainen and Van de Graaf 2018; Obergassel et al. 2019; Blaxekjær et al. 2020).[1]

But it was hard for developing countries to maintain the moral high ground exclusively when there were poorer and less developed countries facing imminent extinction through rising sea levels. These countries, Bangladesh among them, were increasingly seeking to have their voices heard and were critical of all countries, both developed and developing, for not doing more right now to secure the future for those most vulnerable. Yet their voices too rang a bit hollow, given their dependence on the Global North for their current well-being and their futures. While global environmental activists supported the full spectrum of concerns expressed by developing countries and those more vulnerable, they feared that the social protections enshrined within the PA had come to Katowice to die. Developing countries did not ally with them in supporting these social protections for reasons of their own.

In what follows, I bring together the figures of developing countries, now highly differentiated, and developed countries to explore how the negotiations accommodated the tensions and fractures among them. I unfold the negotiation of one of the most contentious tools by which the Paris Agreement was to be implemented, that is, the Nationally Determined Contributions (NDCs). I follow the negotiations from the start of the discussions at COP22 in Marrakech to the finalization of the text at COP24 in Katowice, that is, over seven sessions. Drawing on Annelise Riles's (1999b) contention that the formal, aesthetic elements of international legal practice are as important as its content, I attend to the institutional scaffolding, the proliferation of meetings and documents, and textual strategies, including the use of bureaucratic forms (e.g., the agenda) and metacommentary (e.g., session overviews), to corral discussions, as much as I attend to how issues, such as those of mitigation and equity, which have been dancing around each other throughout this book, fare within the space of negotiations.[2]

Leading up to the Negotiations

Negotiator training underlined the importance of being familiar with the provisional agenda with annotations for upcoming intersessions and the annual COP released before the session started. While scrambling to read and understand the texts before each session, it never dawned on me that the agenda must be coming from somewhere, until, one day, I asked and was informed that in addition to the executive secretary of the UNFCCC and the president of the COP, who together open and close the session and preside over all the meetings, there was a Bureau that set the provisional agenda to be adopted before the commencement of the next session (see Table 1 for the organizational structure of the COP).

The Bureau's task was enormous, providing an agenda for both the intersessional session meeting and the entire COP, with agendas for each of the five conferences under the COP's rubric, including the COP for the Framework

Table 1. Who Organizes the COP?

COP Presidency	UNFCCC Secretariat (see also organizational structure and workflow charts)	Bureau
The Party that hosts the COP (rotates between the five geographic regions of the world and the Small Islands Developing Nations delineated by the UN)	Headed by an executive secretary	Constituted of eleven elected Party representatives from the five geographic regions of the world and the Small Islands Developing Nations delineated by the UN
	Extends its knowhow on conference affairs to host country	
	Contributes some money for the arrangements	
The COP president is usually a high-ranking official of the host country, such as the minister of environment	Provides financial and logistical support to delegations	
	Helps manage the conference venue	Provides advice to the Presidency and Secretariat on process management and reviews Party credentials and the list of non-Party stakeholders seeking accreditation, among other tasks
Makes local arrangements	Serves as rapporteur of the session	
	Produces draft and decision texts	
Contributes the bulk of the money for the arrangements	Manages web portal, broadcasting, and social media	
Takes initiative in leading the negotiations	Translates decision texts into six languages and publishes them on its web portal (unfccc.int)	
Tries to ensure some decisive outcome that carries the name of the host city (e.g., Kyoto Protocol, Paris Agreement, Doha Amendment, Katowice climate package . . .)	Organizes workshops, intersessional sessions, and other meetings in preparation for the COP	Sets the agenda for sessions and intersessional sessions for COP, CMP, CMA, SBSTA, and SBI
	Organizes side events and mandatory events at the COP	

Convention, the CMP for the Kyoto Protocol, the CMA for the Paris Agreement, and the sessions of the subsidiary bodies SBI and SBSTA. Since the CMA was in abeyance at the start, as the PA had yet to be put into effect, the Bureau also provided an agenda for the APA, whose task was to operationalize the PA and put CMA back into the rhythm of an annual session (see Table 2 for the sequence of meetings leading up to the COP, for which the Bureau also provided agendas).

The agenda set for the first-ever APA session held in Bonn in May 2016 during the intersessional session confounded a layperson such as myself, as it seemed to have skipped everything that made the text of the PA so rhetorically exciting, including recognizing the historically marginalized, acknowledging the importance of nonstate actors alongside national sovereignty, granting equal recognition to historical responsibility and contemporary realities, and linking human rights to climate change. However, most of this language was in the Preamble to the PA and quite irrelevant to the task of implementing the articles in the PA, which were by and large focused on accounting of the highest standards.

Table 2. Meetings Leading up to the Annual COP

Intersessional session	Meeting of the supreme, subsidiary, and sometimes some constituted bodies midway through the year (usually June) in Bonn at the headquarters of the UNFCCC for preliminary negotiations. Second in importance to COPs. Decision texts may be drafted there, although they cannot be adopted there.
Pre-COP ministerial meeting	High-ranking officials of a representative group of Parties meet in the weeks leading up to the COP to clarify the main political issues. Usually organized by the host country.
Presession meetings	Some of the constituted bodies and expert groups (the Adaptation Fund, WIM ExCom, Consultative Group of Experts-CGE) have meetings just before the COP. Usually set up by the Secretariat.
Regional group preparatory meeting	Various negotiating groups (AOSIS, LDC, AGN, SIDS, the G-77 and China) get the Secretariat's help organizing regional meetings to prepare for the sessions. Usually the responsibility of the host country.
Pre-COP group coordination meetings	Various negotiating groups (AOSIS, LCD, AGN, SIDS, the G-77 and China) get the Secretariat's help in organizing meetings to coordinate their positions just before the COP. Usually the responsibility of the Secretariat. Other blocs may also hold meetings at this time, but these are not the responsibility of either the host country or the Secretariat.

What follows are the initial agenda items in full. It was revised before being adopted; thereafter, no changes were made during the seven times the APA met between 2016 and 2018 before concluding its work at COP24 in Katowice:

1. Opening of the session.
2. Organizational matters.
3. Further guidance relating to nationally determined contributions referred to in Article 4 of the Paris Agreement.
4. Further guidance in relation to the adaptation communication, including, inter alia, as a component of nationally determined contributions, referred to in Article 7, paragraphs 10 and 11, of the Paris Agreement.
5. Modalities, procedures, and guidelines for the transparency framework for action and support referred to in Article 13 of the Paris Agreement.
6. Matters relating to the global stocktake referred to in Article 14 of the Paris Agreement.
7. Modalities and procedures for the effective operation of the mechanism to facilitate implementation and promote compliance referred to in Article 15 of the Paris Agreement.
8. Preparing for the entry into force of the Paris Agreement and for the convening of the first session of the Conference of Parties serving as the meeting of the Parties to the Paris Agreement.
9. Other matters.
10. Closure of and report on the session.

The Bureau took its directions from the COP decision (1/CP21) that both adopted the PA and provided instructions on how future work on each article of the agreement was to proceed. The work program for mitigation was largely restricted to reporting nationally determined contributions. NDCs were the one mandatory reporting obligation, as indicated by using "shall" within the agreement, which was otherwise dotted by recommendations couched in the language of "should," and was the PA's crowning glory. Starting from even before the Paris Agreement was adopted in 2015, countries were asked to submit what they anticipated to be their contribution to curbing carbon emissions in order to meet the objective of the Convention laid out in Article 2.[3] The qualifier "nationally determined" indicated that nation-states would decide for themselves what they would do for climate action, unlike the KP, which set country targets for emissions reductions.

The initial version of the agenda was not without controversy. After much discussion at the COP opening plenary, the agenda item relating to mitigation was made more expansive (see Table 3 for the work undertaken during the COP). APA agenda item 3 went from "3. Further guidance relating to nationally determined contributions referred to in Article 4 of the Paris Agreement" to being:

3. Further guidance in relation to the mitigation section of decision 1/CO21 on:
   (a) Features of nationally determined contributions, as specified in paragraph 26;
   b) Information to facilitate clarity, transparency and understanding of nationally determined contributions, as specified in paragraph 28;
   (c) Accounting for Parties' nationally determined contributions as specified in paragraph 31.

Adaptation was being considered under APA agenda item 4. However, other issues central to climate action remained "homeless," such as those of climate finance, and were taken up to be facilitated by the co-chairs of the APA contact group, whereas capacity building, technology transfer, and loss and damage were punted to the SBI and SBSTA agendas, already heavy with the preexisting workloads of the Framework Convention, the KP, and other bodies and funds. There were now new pillars of climate action, namely, a transparency framework, stocktaking, and implementation/compliance. These pillars were also to provide the NDCs with the correct orientation for reporting, compliance with reporting, and utilizing the reports to gauge the extent of climate action underway on a regular basis. There was a color-coded work progress tracker to chart the work, stretched across five meetings, on any one of the pillars of climate action. Mitigation was coded a stormy blue. All the decisions taken by APA, SBI, and SBSTA to implement the PA constituted the Paris Agreement Work Programme (PAWP).

While not all Parties and observers took kindly to this agenda, seeing it as reducing the ambition of the PA to that of an auditing company, work on these agenda items began almost immediately, with the formation of a single contact group for APA, which was to report regularly on progress on agenda items 3 through 8 to the CMA and the COP. The contact group was to be headed by two women, one from a developed country (Jo Tyndall of New Zealand) and another from a developing country (Sarah Baashan of Saudi Arabia). The two heads of the APA contact group established consultative groups to carry out the technical work, one for each of the agenda items. Each group was to be headed by two co-facilitators, who were again carefully selected to maintain a balance between developed and developing countries. The consultative group for APA

Table 3. Work Undertaken during the COP

Participants	Week 1 (Technical Work)	Week 2 (High-Level Segment)
Parties	Presession meetings by negotiating blocs to harmonize positions; blocs assign lead negotiators to monitor and shepherd specific issues	Heads of state/ministers/high-ranking officials come to represent Parties (this may even happen at the start of the conference)
	Daily meetings by blocs to monitor negotiations across issues	Hold high-level segment opening meeting with statements by political leaders and observers
	Negotiate text	Attend closed-door meetings to facilitate agreement on major political issues
	Engage in bilateral discussions	Engage in bilateral discussions
	Engage media and civil society	Engage media and civil society
COP (bookends the entire session)	Opening plenary, open to all	Closing plenary, open to all
	Nonadoption of draft rules of procedures (given the lack of agreement on one point of procedure, the rules have never been adopted, although they are considered at every COP, and all other rules are applied. This nonadoption of rules means that all COP decisions must be by consensus rather than by quorum)	Approval of final conclusions and/or adoption of decisions with interventions by Party delegates
	Adoption of agenda, with interventions by Party delegates (an annotated provisional agenda will have been precirculated)	Prepared statements by Parties and observers
	Prepared statements by Parties and observers	
CMP/CMA	Opening plenary that leads on immediately from COP opening plenary, open to all; adoption of agenda for the specific body; nomination and election of a pair of chairs (one from Annex I countries and one from non-Annex countries) to oversee ad hoc bodies, such as APA; other procedural matters brought up	Approval of draft conclusions and decision texts forwarded to the COP Presidency
	Following on opening plenary, chairs hold stocktaking meetings regularly throughout the first week to gauge progress on the agenda items for their bodies (open to observers)	Closing plenary that leads immediately to COP closing plenary, open to all
	Certain agenda items or subitems that need more discussion, perhaps because a decision is expected on it at the current or upcoming session, are highlighted in plenary. They are assigned a pair of facilitators (again from Annex 1 and non-Annex countries) to lead contact group meetings on the topic. Contact group meetings may not be open to observers upon request of any of the Parties, although the first and final contact group meetings are customarily open	

(Continued)

Table 3. Work Undertaken during the COP (continued)

Participants	Week 1 (Technical Work)	Week 2 (High-Level Segment)
	As facilitators of contact groups are only allowed six official informal consultations per contact group, they may request further informal consultations among Parties aside from contact group meetings throughout the week. They must ensure that there is not much conflict of informal meeting times across agenda items to prevent Parties with smaller delegations being denied robust participation in negotiations. Informal informal consultations are not publicly announced and are closed to observers	
	Informal consultations and informal informals may break into bilaterals to facilitate negotiations on sticking points. The president of the COP or chairs of the body may create a small, balanced group of delegates called "Friends of the Chair" to help with these consultations (notes are not maintained of these meetings)	
	CMP/CMA stocktaking meetings throughout the week to gauge progress across each agenda item, open to all	
SBSTA/SBI	Opening plenary that leads on immediately from COP opening plenary, open to all; adoption of agenda for the specific body	Closing plenary with draft conclusions and decision texts forwarded to COP Presidency to be taken up for adoption at the COP closing plenary
	SBI and SBSTA have their standing chairs. If there is an issue on which a decision is imminent, they have the authority to set up informal consultations without need of a contact group. These SB informals have the structure of contact groups, however. They too have stocktaking meetings and can produce draft decision texts. SB informal consultations may not be open to observers upon request of any of the Parties	

Observers (UN entities, intergovernmental bodies, NGOs, activist organizations, all with accreditation)	Dialogue sessions or briefings with the COP president or presidency team and/or members of the Secretariat
	Plenary interventions and formal statements
	Formal statements at the high-level segment opening meeting
	Attendance at as many contact groups and stocktaking meetings as possible; notes taken for widespread distribution
	Daily meetings with civil society constituencies to update on negotiations (the most active ones being RINGO, YOUNGO, Women and Gender constituencies, IPO, BINGO, TUNGO)
	Organization and participation in side events
	Participation in Climate Action Studio/Action Hub
	Organization and participation in exhibits
	Press conferences
	Attendance of gala events by host country
	Requests to COP president, UNFCCC executive secretary, and Party delegates for bilaterals
	Climate march on the Saturday between the two weeks
	Events and actions inside the venue with permission
	Events and actions outside the venue with permission of host country authorities

agenda item 3 was to be headed by two individuals, a woman from a developed country (Gertraud Wollansky from Austria) and a man from a developing country (Sin Liang Cheah from Singapore).

The facilitators were to call meetings of their groups. The work of each group was to produce a text on the relevant agenda item, beginning with an outline with sections and subheadings, then populating these sections with various options suggested by Parties or blocs and continuing to discuss these options until such a time as they had been whittled down to one and the square brackets (suggesting a provisional nature) around words, phrases, and paragraphs removed, at which point the text remaining would be by default what everyone agreed to (see Figure 1, the Matrix of Documents within Negotiation Rooms, on the document types and flows within the negotiations; also see Riles 2000).

I provide this lengthy description to demonstrate the exhaustive ways in which the Secretariat used texts and their production to keep the process on track. At the same time, the organizational setup was routine across all the sessions and showed what fine calibration the Secretariat had to undertake to uphold the UN's principles, such as that of gender equity and respect for national sovereignty, while attending to global power differentials. Although outcomes

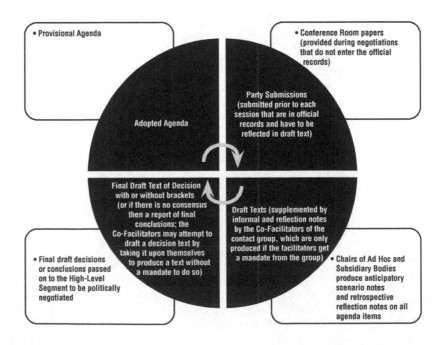

oftentimes felt predetermined—after all, the agenda items had already narrowed the field of possibilities—Party delegates of less powerful countries nonetheless felt that there was enough scope for their perspectives to be heard and
addressed to make engagement in the process worthwhile as a genuine exercise
in geopolitics, even if not always in meaningful environmental action.

The Making of Nationally Determined Contributions (NDCs)

Taking my cue from Zia bhai, I followed the negotiations on APA agenda item
3, "Further guidance related to nationally determined contributions referred
to in Article 4 of the Paris Agreement," which had been expanded to encom-
pass the features of NDCs, what information Parties needed to provide to ensure that their NDCs were transparent and understandable, and, finally, how
accounting for NDCs would work.[4]

Following the changing annotations specifically relating to agenda item 3 in
the provisional agendas from meeting to meeting helped me track prep work
and progress on the issue of NDCs. In the first agenda, dated 7 April 2016 for
APA1 in the May intersessional session in Bonn, the action for this item was
simple: "The APA will be invited to initiate the work referred to in paragraph
13 above" (UNFCCC 2016, April 7).

In the second agenda, dated 1 September 2016 for APA1.2 at COP22, we learn
that Wollansky and Cheah, the co-facilitators of the consultative group for
agenda item 3, initiated work on this item by inviting Parties to submit their
views (UNFCCC 2016, 1 September). Tyndall and Baashan, acting as the APA,
instructed the group to continue its consideration of the agenda item, including Parties' views. In other words, the focus was on determining what Parties
wanted to have included in the draft of the decision text.

In the next agenda, dated 24 February 2017 for APA1.3 at the May session in
Bonn, the annotations for agenda item 3 indicate that there had been a first
pass at a discussion by the Parties after submitting their views, and questions
identified by the Parties as relevant to this item had been listed in an annex to
an informal note by Wollansky and Cheah (UNFCCC 2017, 24 February). Tyndall and Baashan, acting as the APA, had also provided a scenario note on the
progress and sticking points of all the APA agenda items, including the one
on NDCs. Parties were requested to submit further views on agenda item 3.
Wollansky and Cheah requested that the Secretariat organize a roundtable outside of the mandated contact group informal consultations, which would take
the Parties' prior submissions as the basis for an exchange of views. The action
item for agenda item 3 for the upcoming intersessional session was to continue

its consideration of the agenda item, including the Parties' submissions and the views expressed at the roundtable.

By the fourth agenda, dated 25 August 2017 for APA1.4 at COP23, there had been a veritable explosion of texts around agenda item 3 (UNFCCC 2017, 25 August). In addition to the first round of Party submissions, informal notes by Wollansky and Cheah, a second round of Party submissions, and a roundtable where views were also captured, there was now a nonpaper written presumably also by Wollansky and Cheah that captured "convergences, divergences and options" based on the Parties' views in their various submissions but without "omitting, reinterpreting or prejudging Parties' views." Furthermore, the APA requested that the Secretariat organize a second roundtable to take into consideration all previous submissions and the nonpaper. The action for the upcoming session was for the consultative group to continue its consideration of this agenda item. The situation suggested that there was considerable division among the Parties.[5]

In the fifth agenda for APA1.5, dated 26 April 2018 for the intersessional session in Bonn, the annotation is much briefer than in previous iterations, stating that consideration of the agenda item had continued in the previous session and that progress made was captured in another informal note prepared during the session by the co-facilitators, which was a stupendous 180 pages (UNFCCC 2018, 26 April). The discussion on NDCs had congealed around this one text. But the length of the document pointed to deep divides. The action item continued to underline openness to the Parties' submissions.

In the sixth agenda, marked 4 July 2018 for an out-of-turn additional session in Bangkok, APA1.6, which was called to advance the APA's work program, the annotation shows Wollansky and Cheah exercising all available tools at their disposal to bring the Parties together behind a negotiated text (UNFCCC 2018, 4 July). They had provided another informal note, this time thirty-four pages long, to function as a "navigation tool" to supplement the previous informal note of 180 pages.

In the final agenda, dated 12 October 2018 for APA1.7 at COP24, we learn that an additional tool, a joint reflections note by Wollansky and Cheah, was contained in the Bangkok outcome, in the annex to the report on APA1.6 (UNFCCC 2018, 12 October). APA invited this group to continue its considerations on these matters based on the Bangkok outcome and the aforementioned note, "with a view to completing the work thereon at this session." The time for discussion was over, and the future of NDCs was to be decided once and for all in Poland.

Comparing this back and forth in the discussion around the agenda item that dealt with "modalities, procedures and guidelines for the transparency framework for action and support referred to in Article 13 of the Paris Agreement" suggests that they were likely not just slow and divided but also contentious, more so than other agenda items. For instance, APA agenda item 5 was an equally technical issue. However, although there was a solicitation of Parties' views, one request for a roundtable and one for a workshop, with informal notes and even a joint reflections note written by the co-facilitators, there was only one round of each. Furthermore, there was no explicit mention of "convergences and divergences" and no explicit mention of Parties' sensitivities.

Negotiations through the Voices of Parties

Let us pause here to remind ourselves what these texts, textual actions, and their redirection through other texts were for.[6] They were to produce negotiated draft texts on what NDCs were to look like, to be adopted in the form of COP and CMA decisions in 2018, with the decision to be henceforth applicable to all Parties to the Paris Agreement. The decision was to give guidance on what features NDCs should have; what kind of information Parties needed to provide in their NDCs to ensure that there was clarity, transparency, and understanding; and what accounting modalities were to apply to Parties in terms of the targets outlined in their NDCs. To a layperson such as me, this list sounded like the same ask phrased in three different ways to ensure the Parties were providing sufficiently comprehensive, detailed, and tabulated information on their NDCs to make them easy to add up and compare against one another.

Often it wasn't clear that, having negotiated this degree of nuance for the NDCs, the negotiators fully understood the difference among these elements. In the first round of submissions in October 2016, for instance, Brazil collapsed "features" and "information," claiming that "'further guidance' relates to operational aspects of how Parties to the Paris Agreement will communicate their NDCs, not on the NDCs themselves" (UNFCCC 2016, 7 October). In other words, any guidance should only dictate how NDCs should be reported, not what countries should be doing on the ground.

Canada espoused more clarity on the difference among the three, saying that on "features," the mitigation section of NDCs should have "the contributions that each Party was trying to achieve," "the relevant timeframe," and any "associated conditions." For "information," each Party should communicate "up-front information (UFI) necessary for clarity, transparency and understanding of their NDCs" (UNFCCC 2016, 7 October). The reason proffered for this information was to demonstrate that the Party's NDCs were "fair, comprehensive and reflecting its highest possible ambition." Under "accounting," it was the requirement that Parties account for their NDCs, that is, "continuously track its own progress towards its mitigation goals" and "ensure that information is available when needed to report under the transparency framework and facilitate subsequent technical expert reviews."

China, on the other hand, was among the group of countries from the Global South that feared the Convention was being set aside in favor of the Paris Agreement and that attempts were even being made to renegotiate the PA under the guise of discussing how it was to be implemented (UNFCCC 21, 7 October). China felt that this was to strike the principle of CBDR from the final implemented form of the PA and impose a common set of standards for all countries despite historical differences. Even if countries were not to be held to the same standards of mitigation, this process could still impose burdensome demands of reporting on developing countries. Consequently, China stressed exactly these priorities in its first round of submissions.

On features, China said: "The NDCs should be in full accordance with the principles and provisions of the Convention and the relevant provisions of the Paris Agreement, in particular Article 3, 4.4, 4.5, 9, 10 and 11, reflecting common but differentiated responsibilities between developed and developing country Parties" (UNFCCC 2016, 7 October, 19). On information: "The further guidance for the information should be consistent with the nationally determined nature of Parties' contributions, without introducing common format or undue burden on Parties" (20). Finally, on accounting: "The purpose of elaborating guidance for accounting is to develop general and technical guidance for Parties' reference when they are preparing, communicating and implementing their NDCs, with a view to facilitating the transparency and understanding, rather than to impose detailed common accounting rules or transmit Parties' NDCs into a unified form of absolute emission amount" (21). China would have the Convention acknowledged first before accepting anything like what Canada put forward (see Walsh et al. 2011; Eckersley 2020).

In its submission, India further emphasized equity, sustainable development, and poverty, also important within the Convention: "In terms of Article 4(1), recognize that the timeframe for peaking will be longer for developing countries, and will be undertaken on the basis of equity, and in the context of sustainable development and efforts to eradicate poverty" (UNFCCC 2016, 7 October, 33; see also Bidwai 2012; Michaelowa and Michaelowa 2012; Dubash 2013).

And speaking on behalf of LMDC, the bloc that had seized the mantle of Third World solidarity, Iran said: "This means, therefore, that equity and CBDR permeate and must be reflected in all of the NDC components. While there is a common obligation under the Paris Agreement related to prepare, communicate and implement NDCs, the content, features and information of such NDCs must adhere to the principles and provisions of the Convention and reflect the common but differentiated responsibilities between developed and developing country Parties" (UNFCCC 2016, 7 October, 39; see also Halkyer 2021).

Other smaller developing countries struck a more centrist position, moving between underlining the need for a streamlined NDC common for all, as well as attention to common but differentiated responsibilities. For instance, Papua New Guinea, a SIDs nation, insisted that the NDCs should be mitigation focused and not inclusive of the other pillars of climate action, noting that adaptation belonged elsewhere in the reporting framework (UNFCCC 2016, 7 October). Further, the NDCs should be based on domestic sectoral mitigation targets, should be reported in a standard template accepted by all Parties, and should be mandatory from developed countries and those with economies in transition but also encouraged from developing countries, provided they were given support to prepare the NDCs (Betzold 2010; de Águeda Corneloup 2014).

The positions and divisions became clearer over the course of 2017, as evidenced by Party submissions (Parties 2017, 31 May). China unwaveringly struck the same position on CBDR as before. It continued to push for comprehensive NDCs: "The NDCs should include mitigation, adaptation, finance, technology development and transfer and capacity-building, taking into account differentiated obligations of developed and developing country Parties under the Convention and its Paris Agreement" (n.p.). And it sought clear expectations for developed as opposed to developing countries: "Developed country Parties' NDCs should include both ambitious actions and enhanced provision of support to developing country Parties. . . . The extent to which developing country Parties would effectively implement their NDCs will depend on the adequate provision of finance, technology and capacity building support by developed country Parties, recognizing that enhanced support for developing countries will allow for higher ambition in their actions" (n.p.).

There was a discernible shift in position of some of the developed countries as they went from being categorically opposed to differentiation to accepting it. Although the United States was inflexible and became more so under the Trump administration, insisting that it would not accept any "bifurcation" as it narratively recast CBDR in order to demonize it, there was a modulation in Canada's position (Dimitrov 2016). "Canada proposes that these 'NDC types' should be understood as 'examples' or 'ideal-types' of NDCs that are constructed of different 'variables' such as base years, business-as-usual baselines, reference points, measures of intensity, etc. The guidance would then describe the information necessary for CTU of each variable, so that the complete set of guidance would reflect the full diversity of NDCs without constraining any Party's national determination in choosing the appropriate variables for its own NDC" (Parties 2017, May 31, n.p.). In effect, while there would be a standard template, Parties could retain the flexibility to report on those elements of the template that were relevant to them and as they were able.

Bangladesh struck the same position as that of Papua New Guinea in the earlier round of submissions, insisting that the NDCs be mitigation focused but also at the same time advocating for common but differentiated responsibilities (Parties 2017, 31 May). In its submission, Bangladesh focused attention away from the "features" of the NDCs, saying that they had already been determined and that their content would get more direction by clarifying the "information" and "accounting" aspects of NDCs. Bangladesh put emphasis upon quantified information that would allow Parties to consider if their NDCs were fair and ambitious in the light of national circumstances. It is noteworthy that Bangladesh did not use the language of CBDR or RC (respective capabilities) alone, affixing "in the light of national circumstances" to it. This addition, which first showed up in the text of the Paris Agreement, placed emphasis in the present, while CBDR and RC pulled toward the past.

Bangladesh also made a connection between NDCs and the transparency framework being worked out by the workflow of APA5: "There is a clear linkage between information of NDCs and the process of transparency framework of the Paris Agreement." The importance of this would only become apparent to me much later (Parties 2017, 31 May).

The handful of quotations from the first and second round of submissions between November 2016 and May 2017 gives us a sense of the expansive terrain of the issue and the variety of positions taken. Given that these were to be nationally determined reports, one clear political question was: To what extent was such guidance on NDCs to be prescriptive? And if overly prescriptive, would it not infringe on national sovereignty and the voluntary nature of NDCs? This seemed to be a generalized concern shared across the North-South divide (Lawrence and Wong 2017; Hermville et al. 2019).

Another point that came up frequently was whether the guidance was to carry forward the principles of the Convention or entirely subscribe to the PA. And if it was the latter, how would the distinction between the developed and the developing countries in the Convention, upheld by such concepts as CBDR, RC, "sustainable development," and "poverty," be maintained (see Voigt and Ferreira 2016)?

If the principles of the PA applied, to what extent was the historical record of the large carbon footprint of the Global North's economic development going to count? To what extent were the contemporary conditions of the larger of the developing countries going to count? How were the past and the present to be balanced? These concerns were captured in the phraseology of "equity." Those in the Global South protested that even the expectation of producing an NDC of the same level of sophistication as that expected from developed countries put an unfair burden on developing countries and made it seem as if every Party was equally responsible for the problem and its solution (Voigt and Ferreira 2016).

Furthermore, the PA was seen by the Global South to be a comprehensive agreement, whereas the KP might be faulted for being too mitigation centric. As such, ought not the NDCs present the fullest picture of what a Party was doing with respect to climate action, including adaptation, finance, and the other pillars of action? Shouldn't there be links between the negotiations and work on other parts of the PA? Ought not the work to progress apace (Ellis and Moarif 2015; Christiansen, Olhoff, and Dale 2020; Umemiyaa et al. 2021)?

From the Global North's point of view, the concepts of CBDR and RC were modulated by the qualifying phrase within the Paris Agreement, "in light of national circumstances." The phrase attempted to update the geopolitical and economic situation from what prevailed in 1992. How was the present condition of emissions to count, given that some developing countries were now on the list of top-ten emitters? Why couldn't there be a "common template" for NDCs, with flexibilities claimed by Parties rather than built into the template? Why "bifurcation," when everyone was faced by the same reality and imperative to do what they could? Wasn't mitigation the only remaining solution (Friman and Linnér 2008; Blaxekjær and Nielson 2015; Dimitrov 2016)?

These anxieties, questions, and concerns manifest in the paper trail generated by the negotiations. So how was it that this negotiation came to an end with a mere thirteen-page decision, adopted at the same time as decisions on all the other agenda items at Katowice? I provide three possible explanations, first by examining how the text was winnowed down by looking at the interplay of documents within the negotiation room; second through the Third World Network (TWN)'s coverage and commentary, which captures a Global South perspective on what went on within the negotiations; and third from a firsthand observer and participant, our very own Zia bhai, the mitigation person within the Bangladesh delegation and the LDC bloc's lead negotiator on mitigation.

The Race to a Decision Text: Draft Texts

Looking at the interplay between the texts and the negotiation process, we can perhaps better understand how a short decision text on NDCs came to be adopted, despite all the conflict during the discussions. Clearly, it took decisive political will and strong-arming by the Polish Presidency. However, the Presidency didn't take over the negotiation until the second week of the meeting, during the high-level segment involving Party ministers and other high-ranking officials (see Table 3 for the workflow at the COPs). But there had to be some text in place by the end of the first week, because final decisions are rarely produced ex nihilo.

Here we pay attention to the description marshaled by the co-chairs of APA, Tyndall and Baashan, and the co-facilitators of the NDC-related negotiations, Wollansky and Cheah, in their various notes, where they provide a narrative structure to what might otherwise feel like interminable hours in informal consultations iterating and reiterating Party positions without any clear outcomes. They thus shepherded the process toward a manageable final draft of a decision text for use by the COP Presidency in Poland.[7]

In 2016, a short year after the 2015 Paris Agreement, there was little development other than the adoption of the APA agenda and the parceling out of preliminary work in the form of requesting Parties to submit their views and setting up of roundtables for the exchange of views. But unlike what I had thought about the COP in Marrakech earlier, it was still an important year because it allowed the chairs and facilitators both to acknowledge the diversity and points of tension among the Parties but also subtly underline what was acceptable and what was not within the scope of the APA negotiations. For instance, in their scenario note of October 23, 2016, the APA co-chairs Tyndall and Baashan pointed out that the Parties did not yet seem to have a clear-cut understanding of the difference among features, information, and accounting with respect to NDC. "The information communicated by Parties in their submissions revealed that boundaries between the different sub-items of agenda item 3 are not always clear-cut, with some submissions making, for example, similar observations with regard to both features and information" (UNFCCC 2016, 23 October, 4).

While they took cognizance of the singular importance of this being a Party-driven process, being careful to indicate this awareness through statements such as "it would be useful to clarify how to achieve the purposes of NDCs without impacting national sovereignty," they avoided almost all reference to another principle important to many of the Parties involved, CBDR (UNFCCC 2016, 23 October, 4).

The informal note on November 14, 2016, prepared by the co-facilitators Wollansky and Cheah somewhat mitigated the opposition to CBDR, suggesting that this was yet a fluid discursive terrain (UNFCCC 2015, 15 November). Instead of emphasizing the issue of national sovereignty and the Party-driven nature of the process, as the APA co-chairs did in their scenario note, this note placed its emphasis on the fact that Parties did not want to renegotiate the Paris Agreement: "During our discussions under this agenda item, many Parties highlighted that the work should respect the provisions and spirit of the Paris Agreement and should not lead to renegotiating the Paris Agreement and its agreed provisions" (UNFCCC 2016, 15 November, 1). In a way, this was a concession to Parties of the Global South, since limiting the purview of implementation to just what was articulated in the PA and the accompanying COP21 decision text put a check on those Parties of the Global North pressing for new features to be added to existing NDCs to ensure more quantitative and comparable data. Developing countries considered these additions an overreach.

The note also provided a list of questions to guide Parties' submissions on the tripartite issues of the NDCs. It is interesting that the questions almost take the form of the headings and subheadings of the template of a possible draft decision text (UNFCCC 2016, 15 November, 3).

In a follow-up informal note on May 15, 2017, Wollansky and Cheah provided a short document with the headings of "features," "information," and "accounting," with the subheadings "general reflections" and "elements," under which they assimilated Party responses to their questions as text for a draft outline (UNFCCC 2017, 15 May). In their reflections note on October 16, 2017, Tyndall and Baashan, presiding over the entirety of the APA negotiations, acknowledged and thus legitimated these "skeletons" produced by the consultative group as the draft outline for a future decision text developed by the Parties (UNFCCC 2017, 16 October). There was nothing untoward here, but it is important to note that it took slippages between observations and notes, questions posed in a certain format, responses assimilated to the same format, and a recognition of them as legitimate products of negotiations by a higher authority for a draft text to come into existence and the drafting process to get underway.

COP23 in November 2017 saw three informal notes by Wollansky and Cheah that progressively grew from forty-five to 180 pages. The first, dated November 10, 2017, vastly expanded the subsections under the main headings in the skeletal text, with extensive material in the appendix in "clusters" (UNFCCC 2017, 10 November). By the third note on November 13, 2017, the material in the appendix had grown in detail and been migrated to "options" under the main headings and subsections, with much of it in gray blocks, indicating that it might not belong where it was (UNFCCC 2017, 13 November). The final text came with a table of contents, which gave an appearance of order to an otherwise unwieldy text.

In their reflection note on April 10, 2018, Tyndall and Baashan summarized the work that had been done across all the agenda items and indicated the degree of progress or lack thereof within each (UNFCCC 2018, 10 April). They spelled out the sequence of actions remaining to be undertaken for each agenda item for the work of the first week of the COP session to be completed. The summarizing, evaluating, and sequencing is worth quoting in full to help us understand the power of narrative in giving flesh to anemic texts, providing recognition and official standing to the informal work that yet thronged the texts and both commending Party delegates for their work and encouraging them to stay the course. There was even a subtle reference to brinksmanship with the reminder that this was the final year of APA negotiations. They wrote:

> [There has been a] transition from broad conceptual debates to more specific technical discussions and the elaboration of substantive elements. This transition is reflected in the growing maturity of the informal notes on APA agenda items, the stabilization of the "skeletons" or draft outlines used in the notes, and the richness of their content. Key options and critical linkages identified by Parties have become clearer across the whole spectrum of APA work. . . . Therefore it is our judgment that the APA is "on track" to fulfill its mandate but, given the breadth and complexity of the topics being addressed, we need to significantly increase the pace of work. For that, another transition is needed now: from the identified substantive points/ elements to full textual narratives for the proposals and options on the table. Making such a transition should allow Parties to build a basis for negotiations—this is now the single most urgent task, as 2018 is our final year of the negotiations on the Paris Agreement Work Programme (UNFCCC 2018, 10 April, 1).

Amid all the encouragement was still a hint of anxiety that Parties, particularly those of the South, would not respect the textual and narratological efforts to cordon their demands as to what the NDCs should be "about," as indicated by the chairs' language on linkages: "Linkages are a critical part of the big picture and decisive for determining how the whole Paris Agreement system will work together. Given the state of the negotiations and the growing maturity of the substantive material, interlinkages within and across APA agenda items are becoming clearer." While working on one issue, a Party negotiator had to be cognizant of work being advanced in other negotiating rooms, to respect it and synchronize one's position with it. One could not merely privilege what was being discussed and determined within one's own track and grow its scope inordinately (UNFCCC 2018, 10 April, 2).

Such a statement can be seen as a direct rejoinder by Tyndall and Baashan to those Parties within the negotiations who insisted that the NDCs include reporting on all pillars of climate action. As they argued, adaptation and means of implementation (technology and finance being primary among them) should be considered in the NDCs because mitigation did not make much sense without them. With adaptation in the picture, one could immediately see benefits to both in climate efforts reported in the NDCs. One also needed reporting on finance in the NDCs, as only then would one know how much was needed by Parties to undertake mitigation.

While acknowledging the 180-page document that emerged out of COP23, Tyndall and Baashan anticipated that it would be difficult to navigate. And on cue, Wollansky and Cheah provided a document that they called a "navigation tool" in its first iteration on May 7, 2018 (UNFCCC 2018, 7 May). As with all previous notes, this one too was qualified with the obligatory statement that the co-facilitators had prepared the text by their own decision and not through the mandate of the wider contact group and that it did not "cancel or supersede the Informal note by the co-facilitators—final iteration, of November 13, 2017, issued at APA1-4," that is, the 180-page behemoth.

Wollansky and Cheah presented their work as merely "streamlining" the informal notes "by reducing/removing duplication, clustering material—without removing any options from the table." Streamlining indicated a shift from conceptual debates, technical work, substantive elements, textual narratives, and interlinkages, accomplishments that had already been noted by the APA co-chairs' reflection note of April 10, 2018, to editing, that is, hack work, which was the lot of co-facilitators. With these words, the co-facilitators indicated that the text had now moved out of the jurisdiction of Parties into that of officials.

In an informal document dated August 2, 2018, produced by Tyndall and Baashan, it becomes apparent that such tools had replaced informal notes across all the agenda items (UNFCCC 2018, 2 August). Even though at the time of producing them the contact group facilitators had insisted the tools were only additional to prior informal notes, in their advice for the upcoming session the APA chairs urged that "APA 1.6 may wish to consider the tools, in addition to the informal notes contained in the annex to the APA 1.5 conclusions, in its deliberations" (UNFCCC 2018, 2 August, 3). This edging out of notes by tools was made evident in additional instructions that focused exclusively on the navigation tool. The tool provided line numbering for ease of read-ability, prose within it that was considered too general and not operational was put in italics, there were guiding questions in boxes for each agenda sub-item, and proposals that did not appear to be in the appropriate location within the tool were highlighted to make them movable to elsewhere in the text.

This work was put together through discussions that had been occurring under the auspices of, first, the Subsidiary Body on Implementation (SBI), on public registries to make the information from NDCs widely available and common time frames on how often NDCs were going to be submitted; and second, the joint discussion under the Subsidiary Body on Scientific and Tech-nological Advice (SBSTA) and SBI on response measures, or the adverse im-pacts of mitigation policies upon Parties' economies. Pulling all this work together through interlinkages made one realize how much work had actually been achieved on this topic.

In a final reflection note on the textual products of all the agenda items, issued on October 15, 2018, by Tyndall and Baashan and written in concert with the chairs of SBI and SBSTA, they expressed alarm that Parties had not yet decided between what was to remain in the decision texts and what was to go into the appendix as technical guidance (UNFCCC 2018, 15 October). That is, the texts were as yet too long and lacked a clear indication of what was pre-scriptive, that is, required, and what was informative. They now took com-mand of the process in its final lap to bring texts to "maturity," indicating the heavy hand that they were going to employ in shaping the decision texts that were to be sent on to the high-level segment of the COPs. They wrote:

> Our textual proposals for this session (presented as addenda to this note) have sought to bring all PAWP [Paris Agreement Working Programme, which brought together all conclusions from APA, SBI, and SBSTA] items to a comparable state of maturity and readiness by moving towards draft legal text wherever possible, streamlining the text, collapsing closely related and similar options, looking for possible convergence and suggesting possible solutions while ensuring that the broad options put forward by Parties are clearly delineated so as to assist with the choices that need to be made (UNFCCC 2018, 15 October, 5).

And they went further by doing away with some options entirely, particu-larly those asking for no texts to be provided on particular issues: "In the tex-tual proposals attached to this joint reflection note, however, the APA Co-Chairs have elected to dispense with 'no text' references. Instead, differences in Party views are reflected by the use of square brackets around paragraphs or sections.

Readers should interpret such square bracketed text as implying that Parties will need to reach consensus on any text to be included" (UNFCCC 2018, 15 October, 6). At the same time, the note also provided a way to defer some difficult decisions to work around the mandate to be done with deliberating over the means of implementation of the PA by 2018. "Parties also need to identify any follow-up work of a technical nature that may be needed after 2018 and find effective and practical avenues for such work to be conducted."

The first official draft text on agenda item 3 appeared on December 5, 2018, stated under the heading "Features of nationally determined contributions, as specified in paragraph 26,"

1. *Notes* that features of nationally determined contributions are outlined in the relevant provisions of the Paris Agreement;
2. *Decides* to continue its consideration of further guidance on features of nationally determined contributions at its seventh session (2024) (UNFCCC 2018, 5 December).

The Parties had decided to stop negotiating over features relating to the NDCs, deferring it to 2024, while focusing on information and accounting for the decision text to be adopted.

The second official draft text, dated December 7, 2018, is noteworthy because Tyndall and Baashan provided a note in the preface suggesting that the discussions still remained contentious (UNFCCC 2018, 7 December). They indicated that they had sought the help of yet another "tool" by which to arrive at consensus, that of the "landing zone." "Parties need to continue working intensively on the shaded sections, both for ICTU [Information to facilitate clarity, transparency and understanding] and accounting today (7 December 2018). These sections represent our judgment of an acceptable landing zone, but the language could be improved. We will forward a clean version, take on board improvements developed today, to the COP President tomorrow, on 8 December 2018" (UNFCCC 2018, 7 December, 1). As it turned out, Tyndall and Baashan's landing zone was to be the last word on the deliberations. A final iteration of the landing zone document went forward to the COP Presidency to be tackled through the political process.

With all the texts on the various agenda items of the APA forwarded to the Presidency, the work of APA was completed. The process ended with an elegant decision text, Decision 4 under the auspices of CMA 1, which finally resumed its operation as the work of APA concluded (UNFCCC 2019, 19 March). The text had a short but pointed preamble, emphasizing the importance of attending to the needs of developing countries and of providing them support to enable them to achieve ambitious NDCs, before moving on to the specifics of what information was necessary to "facilitate clarity, transparency and under-standing," which had its own acronym, ICTU, and the accounting called for, with technical lists of what could be included in these in the appendix. The main decision text ended with the decision to move discussion of features to 2024. The text iterated several times that Parties could choose to include ad-aptation information in their NDCs to show mitigation co-benefits. While it is hard to say who came away as a clear victor in this process, developed coun-tries, developing countries, or LDCs and SIDs, it is noteworthy that there was still no mention of the Convention or of CBDR and RC in this decision.

Negotiations through the Eyes of Activists

The TWN was one of the expert organizations that provided support to coun-tries and was known for being explicitly Third Worldist in its politics (see Wil-liams 2005). The group supported the larger developing countries (such as China and India) most explicitly through the LMDC bloc. In addition, they also provided a running commentary on the sessions through regular news-letters circulated before, during, and after the meetings. The TWN's analyses underline my point that the process was political through and through, de-spite the evocation and mobilization of the language of technical work or the neutral language of streamlining and bringing texts to maturity. At the same time, their newsletters were not above elisions and appropriations of their own, which complicate and nuance South-South politics.

On January 9, 2016, Meena Raman of TWN, one of the central experts within climate negotiations (introduced earlier) wrote a summary report on the negotiating objectives of developing countries, which are by now well known to us. This report, first published in *Economic and Political Weekly*, was later republished on TWN's own website (Raman 2016):

> They [developing countries] wanted to (a) defend the Convention and
> not let it be changed or subverted; (b) ensure that the Agreement is
> non-mitigation centric with all issues (including adaptation, loss and
> damage, finance and technology, besides mitigation) addressed and
> in a balanced manner; (c) ensure differentiation in all aspects be
> reflected, with the principles of equity and common but differentiated
> responsibilities and respective capabilities; (d) ensure that developed
> countries enhance the provision of "finance and technology transfer"
> (f) ensure that "loss and damage" is recognized as a separate pillar
> apart from adaptation and for (g) legally binding provisions, especially
> on the developed countries.

Speaking about the United States and the Umbrella Group (Australia, Canada, Iceland, Japan, New Zealand, Norway, the Russian Federation, and the Ukraine), Raman said, "They mounted an onslaught on the Convention, seeking to weaken the provisions and their obligations; redefine differentiation so as to blur the different obligations of developed and developing countries; and a legal 'hybrid' (in terms of what clauses are and are not legally binding), mainly to suit the US administration's relations with the US Congress which is hostile to the climate change issue" (Raman 2016).

Over the course of TWN's coverage of sessions between 2016 and 2018, we see how the developing-country objectives waxed and waned within the climate negotiations. From early on, despite the fact that developing countries made CBDR and RC central within their Party submissions and options, those exact words were avoided in any written text by the co-facilitators and co-chairs (as we saw earlier). TWN notes how the principle went from being ignored to being mocked by some Parties. In the Bonn intersessional session in 2017, for instance, Switzerland referred to it as "confusion, bewilderness [*sic*], delusion and rolling back" in its intervention, saying that it would refrain from stating the acronym for the four terms (Ajit 2017, 5).

The efforts to leave out differentiation on grounds that it had already been negotiated and was a thing of the past or that it was a political issue were repeatedly called out by countries such as India, whose delegation members are quoted by TWN as saying in COP 22 in 2016 that "terming differentiation as a political issue is a nonstarter. Whether we call it political or nonpolitical or technical or non-technical, the fact is the PA is based on equity and CBDR and each element is informed by it. When we discuss how to operationalize the elements, we have to operationalize differentiation also. We cannot set this aside" (Ajit 2016, 2).

Another issue brought up repeatedly by developing countries but given short shrift in the discussions and texts was the comprehensiveness of the NDCs. At the Bonn intersessional session in 2016, TWN reported that there was a fight underway over whether Article 3 or Article 4 of the PA was going to inform PA, with Article 3 seeking to make the NDCs more comprehensive and Article 4 projecting NDCs as mitigation centric (Bose and Raman 2016).

In the TWN newsletters for the next intersessional session in Bonn in 2017, we learn that developed countries and "some developing countries" were pushing for a mitigation-centric NDC with a common template, despite the fact that this would compromise the comprehensiveness of the NDCs and the principle of differentiation within the Convention (Raman and Ajit 2017). We are told that those developing countries that gave support to developed ones were from AILAC, LDC, and AOSIS. This tallies with what we found in the

Party submissions on APA agenda item 3 sketched earlier. The further insinuation by TWN was that smaller developing countries were being redirected away from the G77 and China by developed countries for their own interests.

Earlier, I had mentioned how certain issues were considered to have been rendered homeless by the limited agenda of the APA. We learn in full what these issues were in TWN's coverage of COP22 in 2016 (Ajit and Raman 2016). They included common timeframes of NDCs, how existing NDCs were to be adjusted in light of the negotiations, implementation of response measures, recognition of adaptation by developing countries, guidance to existing funding entities on how they were to serve the agreement, a new collective finance goal, communication of publicly raised financing by developed countries, and guidance on education, training, and public awareness. These were very much the issues that developing countries were agitating for but that were being held hostage by developed countries that first sought common and shared commitments to emission reductions.

By 2018, the focus had swung away from CBDR entirely. Instead, the focus was on the issue of how the NDCs would be possible without finance. At COP23 in 2017, the young female Chinese delegate provided a poetic description of the ways in which the NDCs were shaping up to be only mitigation centric. "At this stage," she said, "there is only flesh and therefore, there is the need for the bones and the blood" (Bomzan and Raman 2017, 3). China suggested that finance could be thought of as the blood that was necessary to bring the entire text together. Switzerland too struck a more conciliatory tone in its comments, perhaps modulated by the bloc it represented. Speaking on behalf of the Environmental Integrity Group (EIG), it called on Parties "to address the full scope of the agreed mandate on mitigation, adaptation, support and transparency" (3). On differentiation, it said "different situations and circumstances had to be taken into account" (3). In a way, the emphasis on finance would allow differentiation to be addressed by other means (Roberts and Weikmans 2017; Weikmans and Roberts 2019).

TWN coverage makes clear the extent to which the drafting process was viewed with suspicion by developing countries. As early as November 2016, Bolivia was quoted as claiming that Wollansky and Cheah had circulated documents that did not reflect all Parties' views (Ajit 2016, 11 November). In further newsletters, it points out that in addition to the circulation of unrepresentative texts, there was also concern that Parties were not reporting back decisions and progress within informal informals of the contact groups. And finally, some Parties complained that the way co-facilitators were distinguishing technical work from political issues was very arbitrary. In Bangkok in September 2018, there was serious pushback against the manner in which some texts, notably the navigation tool, was numbered and some not, notably the 180-page informal note. Although the explanation given was that it was to ease the navigation of a long text, Parties complained that leaving some text over others unnumbered effectively rendered them "status-less" (Raman and Bose 2018). In effect, TWN was calling out the neutral-sounding acts of streamlining and bringing texts to maturity being undertaken by contact group co-facilitators and APA chairs as acts of political interventions into negotiations.

Openly privileging the perspective of Parties from the Global South, TWN reported: "Further discussions gave way to huddles, and in some of those, emotions ran high, where according to sources, the co-facilitators from Singapore and Italy [sic] were seen to have 'pressured some Parties' to hand over the mandate of drafting the text to them, where they promised to 'weave magic' in relation to resolving the differences among Parties. Developing countries held their ground to ensure a level-playing field and said that if Parties' views were not being allowed to be captured in texts, the starting point for negotiations was clearly, still far away" (Raman and Bose 2018). This was just a short month before the Paris Rulebook would be adopted in COP24 in 2018.

TWN remarked that the discussions on NDCs were particularly difficult compared to talks on other agenda items. At COP23 in 2017, TWN pointed out that while there was much obstruction to having a comprehensive NDC within the informal consultations, this comprehensive approach was taken for granted within the negotiations over the transparency framework.[8] It suggested that battle lines had been drawn within the negotiations over NDCs.

Interestingly, at no point did any of the developing Parties threaten to leave the process. At the intersessional session in Bonn in 2018, the Egyptian spokesperson for G77 and China was quoted by TWN as saying that "'the final stretch of operationalizing the PA will be a challenging one' as 'there are varying and often divergent views on many of the Agreement's provisions.' 'This is normal,' it reported, adding that 'we can assure you that we have and will continue to come to the table with an open mind and all willingness to reach accommodations. If this is the prevailing sentiment and attitude, we have every reason to hope that whatever obstacles we may meet on the road to Katowice will be overcome,' Egypt said in conclusion" (Bomzan 2018).

At COP24 in 2018, TWN provides its own reading of how the final draft of the full decision emerged. On December 6, Tyndall and Baashan informed Parties the time had come for them to produce their "own proposals," to help the process forward, and that based on "where possible landing zones are emerging or may lie," they would produce the next iterations of the negotiating texts "under their own responsibility," which would be made available to Parties on December 7 for their consideration. They stressed that they "cannot reflect all views of Parties in the document" but will "ensure that what is proposed is a balance of interests."

As we know from following the negotiations from the view of its draft texts, the revised landing zone document was to become the final decision, as happened with other agenda items as well. The TWN, of course, saw these actions by Tyndall and Baashan to be presumptive, if not an abrupt abridgement of negotiations. As an activist group, it could say this and more, ventriloquizing developing Party critiques while Party delegates maintained "an open mind and all willingness to reach accommodations." Such ventriloquizing was well understood and appreciated within the process, which saw observers as not only serving as "the eyes and ears of the world on the process" but also "saying what could not be said" within it.

On December 10, Michal Kurtyka, the secretary of state for Poland's Ministry of Environment, who was leading the COP 24 Presidency, said that "a significant amount of work remains to be done within the high-level segment to secure a balanced outcome" on the PAWP and outlined the steps to reach the final outcome to be adopted on December 14. New draft texts with as few options as possible (that is, undecided elements) would be presented and whittled down to zero by Parties led by selected pairs of ministers who had been identified to work on "solving outstanding political issues" that could not be resolved at the technical level. The ministers were free to use all possible tools, including open-ended consultations or meetings with individual Parties, to advance their work. And the CMA adopted final decisions for the implementation of the PA late at night on Saturday, December 15, a day later than the meeting had been scheduled to close.

In their detailed analysis of the full decision text, TWN shows how the decision states that the guidance on information necessary for clarity, transparency, and understanding is without prejudice to the inclusion of components other than mitigation (Raman 2018, 19 December). This is viewed by developing countries as a major win on expanding the scope of NDCs, as it means that countries are encouraged to report on more than mitigation alone. In the transparency framework decision, flexibility is provided for developing countries that need it in the light of their capacities, and this too is a partial acceptance of differentiation. As regards the decision on finance, that is, on the ex-ante information to be provided on the projected levels of public finance for developing countries under Article 9.5, developed countries are requested to provide the information starting in 2020 (Bose and Raman 2018). On setting a new collective quantified goal on finance, Parties agreed to initiate deliberations on the goal from a floor of US$100 billion per year, also in November 2020. In regards to the decisions adopted on finance in Katowice, according to a senior developing country negotiator who spoke to TWN, "there were wins and there were losses for developing countries," adding, "this is the nature of negotiations." However, on the whole, according to the negotiator, "developing countries got more wins than losses in the finance related decisions" (Bose and Raman 2018).

In the Plenary of the Joint Closing of COP and CMA, when it had become amply clear that there would be no give on CBDR, there was mixed reaction. Gary Thesiera, whom I have introduced earlier, a member of the Malaysian delegation who was previously the spokesperson for the G-77 and China within APA and was now the spokesperson for the LMDC, said that the bloc, although pained by some of the provisions, would work constructively to move forward in a spirit of compromise and in interpreting the decision in the context of preserving its fundamental elements and ethos. With his emphasis on "fundamental elements and ethos," he was announcing that CBDR was not dead by any means (Raman 2018, 18 December). TWN described the EU as claiming that "the decisions preserve the notion of 'contemporary differentiation,' recognizing the economic and social evolution of Parties" (Raman 2018, 18 December, 5). Differentiation would no longer refer back to the historical past or even to what prevailed in 1992 but rather to what existed as differences between countries at present.

Explaining why developing countries supported the PA despite its many flaws in January 2016, Meena Raman of TWN wrote in the Indian leftist magazine *Economic and Political Weekly*: "True, the Paris Agreement also means that big pressures will be put on developing countries, and especially the emerging economies, to do much more on their climate actions, including mitigation. But these enhanced actions need to be taken, given the crisis of climate change that very seriously affect developing countries themselves" (Raman 2016). At COP23 in 2017, TWN began its reporting by reminding the reader of the damage wrought by rising global temperatures by pointing to devastating monsoons and floods in Asia; ravaging droughts, landslides, and floods in Africa; hurricanes, cyclones, floods, and forest fires in the Americas; and heat waves in Europe. TWN had its political commitments to the Global South, particularly the larger developing countries that ultimately held the key to Global South and South-South politics, but it was also committed to the central issue of combating climate change at the heart of the process.

Bangladeshi Eyes on the NDCs

After COP24 in Katowice wrapped up, I went home clutching the text, wondering how a decision of only a few pages would suffice to provide the guidelines countries sought. The following summer in Dhaka in 2019 and more recently over Zoom in July 2020, I followed up with Zia bhai, now ensconced in his office in the Bangladesh Department of Environment within the Ministry of Environment, Forest, and Climate Change. His office was a large, air-conditioned, impersonal room, hinting at how little time he spent in it. After all, he was often away at international conferences and workshops. I spent many happy hours in his office working at my computer or dozing, waiting for him at appointed meeting times while he was stuck in traffic or at a meeting at the ministry. Contrite for making me wait, he invariably shared his lunch or tea with me afterward.

I asked him for his opinion of the outcome on mitigation in Katowice. He readily confessed that the negotiations on the issue were a failure and that the text agreed upon was going to be ineffectual. "What happened in that room?" I asked.

"Everyone dug in their heels," he answered. "They couldn't let go of their positions. LMDC produced one obstruction after another. EU and LDCs, with whom we are together on this issue, got more and more frustrated. We needed a decision on what NDCs should look like, and quickly, because the INDCs we had submitted back in 2015 needed to be updated, made more ambitious, and we needed to have information coming in quickly from all the big countries so that we were prepared for the first stocktake scheduled for 2023. But there was no budging anyone. But," Zia bhai continued cheerfully, "it doesn't matter because our objectives were achieved elsewhere in the process."

"Where?" I asked.

"In the negotiations over Apa 5," making a note to himself to congratulate that group for arriving at a wonderful decision text after tricky negotiations. "Apa 5" was the APA agenda item to produce the transparency framework by which all the work for the PA would be undertaken.

"How will that help? What makes it wonderful?" I asked him.

"They provide precisely all the information that we could have gotten from NDCs but now can get from BTR [Biennial Transparency Reports], and that will come more frequently than NDCs, every two years." Meeting the obligations for one will predefine the features of the other well ahead of the 2024 date set to discuss the features of the NDCs.

I asked if Zia bhai if he thought that the LMDC was just in it to obstruct, like the United States. He demurred. "They too want to tackle this problem; they are just a bit *bhitu* [cowardly] about taking on such a large responsibility." I asked why it was that they felt that LDCs and SIDs were so easy to manipulate. He chuckled and reminded me that although they acted as though LDCs and SIDs had just recently been bought off by developed countries, in actuality, the two groups were protected within the Convention itself; that is, they had a special status that was secure for now, whereas the Convention had always expected that developing countries would change and join developed countries. "Developing countries keep trying to remove all the difference amongst us to take over all the provisions made for smaller, vulnerable countries." In the process, the larger developing countries and their supporters denied any independence of thought or integrity of action to smaller, less-developed countries. Yet, Zia bhai reminded me, they were not above appropriating issues of concern, such as loss and damage, the idea for which originated from small island nations and least developed countries and treating it as their own.

"The fight is not yet over," Zia bhai concluded cheerfully. I took him to be referring to the deferred discussions on specific aspects of the NDCs. Or he could be talking of the issues of finance and loss and damage, near and dear to developing countries but yet to take center stage within the negotiations. Or maybe of the hot but contentious issue of the market mechanisms by which Parties were going to undertake climate mitigation, referred to in PA article 6, barely discussed under the rubric of the PAWP, unsuccessful in Madrid in 2019 and now on the agenda for Glasgow, scheduled for November 2021.

The question for me was whether South-South relations would continue to enact what was fast becoming a somewhat repetitive polarity between the issue of differentiation and equity for developing countries and the urgent need for mitigation by all or whether with decision texts in hand, one text would compel another and so on till something shifted in Parties' positions. Such was the power of texts within the climate process to keep Parties and process moving and constantly modulating in relation to one another.

7

The House of Loss and Damage

Most of the articles in the Paris Agreement are as expected when it comes to dealing with the climate, but for one. Article 8, subsection 1 reads: "Parties recognize the importance of averting, minimizing, and addressing loss and damage associated with the adverse effects of climate change, including extreme weather events and slow onset events, and the role of sustainable development in reducing the risk of loss and damage." The "loss and damage article," as it is informally called, stands out because it doesn't spell out any actions to be taken other than cognizance of possible climate impacts.

Each of the other pillars of climate action is backed by consensus among the various constituents of the process as to what it involves, but there isn't a shared understanding of what this article is "about." Loss and damage may be alternatively an acknowledgment of suffering, a sharing of best practices, a pressure tactic, an insurance scheme, and many things besides, depending on who is speaking. We know merely that, according to the IPCC, "loss" refers to that which is irrevocably lost and "damage" to that which is seriously impacted but still retains the capacity to recover.[1]

Having previously explored how the issue of equity, in its many forms, tussles with the urgency of mitigation across COPs, we turn to loss and damage in this final chapter, as it offers us a different cross-section of these issues. Loss and damage emerged in response to the failure of developed countries to take the lead in mitigating. Support for it, however, became more broad based when loss and damage appeared to provide a means to make finance more central as a way to address inequity. After all wouldn't developed countries prefer to pay into a fund to disburse monies as and when countries confront extreme or chronic impacts of climate change than be found liable to compensate for such damage? While G-77 and China came to adopt loss and damage as an issue on which it had a common position, it should not be assumed that there was unilateral support for it within the bloc. Loss and damage risked pulling developing countries into a mitigation-centric approach to climate change, as has been the case for the LDCs, as it previously was with the AOSIS, while making the larger developing countries within it vulnerable to loss and damage–related litigation.

In this chapter, I explore how the newest pillar of climate action, loss and damage, emerged historically and how developed countries within the process attempted to control the evolution of the concept. I show how it ramified in many different and unexpected directions. The chapter takes us to Bangladesh, to both my primary field site and the national context, to understand how the country attempted to create leverage for itself within this process by pursuing the potential within loss and damage. It ends with speculating on a possible need for this article within the agreement beyond that claimed by its supporters.

While there is much that has been written about loss and damage from the perspective of science (Huggel et al. 2013; Huggel et al. 2015; James et al. 2019), policy (Huggel et al. 2015; Ohdedar 2016; Kreienkemp and Vanhala 2017; Calliari 2018), law and litigation (Verheyen 2015; Adelman 2016; Lusk 2017), and social science (Warner et al. 2012; Huq, Roberts, and Fenton 2013; Nishat et al. 2013), in this chapter I restrict myself to exploring how loss and damage appeared from Bangladesh's perspective while recognizing the polyphony within Bangladesh's discourse. My interest is to draw out how a country ventured into supporting an issue that could earn it the ire of its sponsors and of the larger developing countries in the blocs with which it was associated, and whose speech was suspect for that reason. I posit that its support of loss and damage didn't just derive from Bangladesh's deployment of the power of moral authority, a strategy it was known to practice along with other vulnerable countries, such as in the instance of the Climate Vulnerable Forum. Bangladesh's support for loss and damage, or rather its mode of coming to support the issue without knowing where such an action might lead, pointed to its practice of "weak ontology," the act of making small, uncertain leaps it affirmed for itself or for which it sought the affirmation of similar vulnerable countries (White 2000).

A 2013 blog post by Saleem on the IIED website provides a vivid image of loss and damage. Saleem refers to loss and damage as "a house with many rooms" (Huq 2013). In a manner reminiscent of the memory palaces of Greek

and Roman rhetoric (Yates 2011), the rooms contain the different effects of climate change. People are invited to enter Saleem's house of loss and damage.
It is indicated to the visitors that the house contains rooms with doors titled
"human cost" and "economic cost." While acknowledging that there is substantial presence in the room titled "human cost" (it is "inevitable"), Saleem
states that nothing can be done about that. In so far as there is no clarity on
who pays for human cost, the door to it remains firmly shut. Instead, he invites
his visitors into the room titled "economic cost," and there he further invites
them into annexes titled "research to minimize costs," "sharing knowledge,"
"insurance," "solidarity payments," and "risk retention."

We are told that some already visit this house, particularly "poor countries."
As yet, "rich countries" refuse to come in, but Saleem assures them that the
doors to "liability" and "compensation" remain firmly closed because of another door leading to them labeled "attribution." Without clear causal links
between the actions of rich countries and natural disasters and weather events
in poor countries, there is no reason to enter this room. Its door remains closed,
like the door to "human cost."

The blog post was written in 2013, the year that the issue of loss and damage was brought up forcefully at COP19 in Warsaw. It was an endorsement
of the issue of loss and damage by someone known to be an unofficial insider within the climate negotiation process and trusted by developed countries. It tacitly assured them that the darker aspects of loss and damage, as
a pressure tactic or financial liability, were under wraps. And it subsumed
the heterogeneity among countries by putting countries either in the camp
of the rich or the poor, making this a moral issue rather than an explicitly
political one.

If one looks up "loss and damage" on the UNFCCC website, we get perhaps one of the clearest timelines on any issue within this otherwise tortuous
platform. Although it is useful, I deliberately avoid replicating this chronol-
ogy, because it hints at an institutional effort to corral understanding of the
issue by putting it within the scope of adaptation, whereas most of the fight
within the process has been to accord loss and damage independent status as
the third pillar of climate action, alongside mitigation and adaptation.[2] After
all, it existed because mitigation had failed and because adaptation would not
be able to cope with the scope of changes that people would have to accommodate (Calliari, Serdeczny, and Vanhala 2020). Instead, I follow Saleem's blog
post, exploring it one room at a time, to avoid imposing too much coherence
upon an issue that evolved in fits and starts and that even now has an unset-
tled quality.

Loss and Damage as Human Cost

We'll start with the door leading to the room titled "human cost." One can well imagine what lies beyond it. Every day the media reports on human suffering from seemingly natural disasters, ranging from cyclones, floods, and desertification to insect infestation of crops. With mitigation nowhere close to the levels needed to curb carbon emissions to offset these events, it pointed to the lack of collective will to undertake necessary mitigation. "Human cost" also pointed to the limits of human plasticity and adaptation. Despite efforts to ratchet up adaptation funds, this ratchet had come at the loss of development funds, with a tradeoff observable between undertaking development and undertaking adaptation by developing countries (Hulme, Saffron, and Dessai 2011; Weikmans et al. 2017). Given these limits within the existing scope of mitigation and adaptation, one might say that the door of loss and damage into "human cost" would be an indictment of the continued failure of the climate negotiation process to deliver meaningful results. Many commentators have remarked that the resistance to loss and damage from within the process comes from not wanting to acknowledge this failure (Vanhala and Hestbaek 2016).

For the moment, I take at face value Saleem's contention that human cost was beyond calculation and that it was unclear who should pay for it, making it necessary to close that door, even as my heart sank as I absorbed what the words "human costs," "beyond calculation," "who should pay for it," and "firmly closed" truly meant. Instead, I present a vignette from my fieldwork among farmers living itinerant lives on the shifting sands of the Brahmaputra-Jamuna River. It provides a different picture of engagement with loss and damage than one of endless suffering and its incalculability. It perversely gets to the *jouissance* that I sometimes sense in conversations about loss and damage within the negotiations, in which the issue was raised to rattle developed countries, who were otherwise secure in their moral stance on climate change and sanguine about their contributions to the common good, having made peace with their violent histories of expropriation via colonialism, even if those who suffered these histories were not able to feel that same peace.

At the conclusion of COP21 in Paris, where loss and damage had a major win, in that it was included as an article within the Paris Agreement, I headed to Bangladesh to the field site that had jumpstarted my interest in climate in the first place. I came in winter and was treated to a range of stories people had saved up for me of their experiences of floods in the summer earlier that year. One story in particular provided insight into the extent to which loss and damage was bound up with the flow of life.

Nur Hashem recounted to me how he returned to his home on the silt island after a hard day laboring at a weaving factory on the mainland to find his
village collapsing into the river. He searched frantically for his family, only to
learn that they had fled earlier with neighbors. As he tried to recover what he
could of his fast-eroding household, he suddenly experienced a feeling of complete futility. Rather than succumb to it, he leaped up and started kicking at
his house's walls, helping them collapse into the water. "Here you go, take that," he
shouted. "Go." Momentarily exhilarated, he left in search of his family. Nur
Hashem was spent. Exhausted by his work as a weaver, which is known to take
a toll on the body, he was doubly exhausted by the labor of living beside an
active river. He was throwing in the towel, but in such a way as to transfigure
his feeling of hopelessness into that of elation, albeit temporarily. Within the
vast, multistranded accounting undergirding global climate politics, would loss
and damage be able to account for Nur Hashem's self-depletion, loss of home,
and philosophical equanimity?

Loss and Damage as Economic Cost

Before we make any ready assumption that we know what loss and damage is
"about," what it "means," let us go where Saleem beckons, into the room named
"economic cost," where he asks that we add doors to annexes titled "research
to minimize costs," "sharing knowledge," "insurance," "solidarity payments,"
and "risk retention." As mentioned before, Saleem wrote his blog post in 2013.
Later that year there was a major win for loss and damage at COP19, where it
was decided, after intense and contentious negotiations, that there would be
an independent mechanism for loss and damage, the Warsaw International
Mechanism on Loss and Damage, referred to as WIM.[3] This was two years
before loss and damage got its own article within the Paris Agreement, indicating that the concept had arrived on the scene, entrenching itself within the
negotiations.

During COP24 in Madrid in 2019, Saleem gave me what he called a "potted
history of loss and damage," specifically of the arrival of WIM. He said that
the issue had been raised by the Solomon Islands in 1991. It turns out it was
actually Vanuatu that had proposed an insurance scheme to offset loss and
damage in the early years, but their proposal fell on deaf ears, because the
process was mostly inclined toward mitigation at the start.

It is noteworthy that the Framework Convention mentions damage, both
serious and/or irreversible. In Article 3, it is stated as a matter of principle that
"the Parties should take precautionary measures to anticipate, prevent or minimize the causes of climate change and mitigate its adverse effects. Where

there are threats of serious or irreversible damage, lack of full scientific certainty should not be used as a reason for postponing such measures." It would seem, therefore, that loss and damage was present right from the beginning of the negotiation process. As we know from earlier chapters, if there was relevant wording within a foundational text, it provided the impetus for further action, even if the action was only toward greater discursive elaboration till it burst upon the world.

In the early 1990s, Saleem thought loss and damage was an islands' issue, while the issue for least developed countries was adaptation. Despite not winning the ear of developed countries or the support of the developing and least developed countries, the islands persisted with this issue. The Bali Action Plan decided at COP11 in 2007 mentions the phrase "loss and damage" under enhanced adaptation in the following manner: "Enhanced action on adaptation, including, inter alia, consideration of . . . (ii) Disaster reduction strategies and means to address loss and damage associated with climate change impacts in developing countries that are particularly vulnerable to the adverse effects of climate change" (Huq, Roberts, and Fenton 2013).

In the Cancun Agreements, which emerged out of COP16 in 2010, loss and damage got more text under "enhanced action on adaptation" (Warner 2012). The process "recognizes the need to strengthen international cooperation and expertise in order to understand and reduce loss and damage associated with the adverse effects of climate change, including impacts related to extreme weather events and slow onset events" and further on "decides to hereby establish a work program in order to consider, including through workshops and expert meetings, as appropriate, approaches to address loss and damage associated with climate change impacts in developing countries that are particularly vulnerable to the adverse effects of climate change." It's noteworthy and salient for loss and damage's advocates that the distinction was made between extreme and slow-onset events.[4]

Returning to Saleem's potted history, we learn that it was only after COP17 the following year in Durban that negotiators for the LDCs congregated in Dhaka to be further educated on the issue. Bill Hare of Climate Analytics came to Dhaka, along with Koko Warner, who was at that time the head of the Environmental Migration, Social Vulnerability, and Adaptation Section at the United Nations University. Later she would become the point person between the Secretariat and the WIM (Hare et al. 2011). It was in Dhaka that the LDC negotiators realized for the first time that there were "scientific limits to adaptation" and that "loss and damage was already happening in their countries." Although these were canned phrases, they are nonetheless telling—a moment of mutual acknowledgment of suffering across blocs within the process.

After that point, the LDCs and AGN fell in line with the AOSIS on the issue of loss and damage.

As we have seen in earlier chapters, developing countries were determined that common but differentiated responsibilities (CBDR) should inform climate negotiations because the presence of CBDR assured them that their historical national circumstances would be taken into consideration. That way they wouldn't be committing to more than they could provide, leaving their own citizenry with the short end of the stick. With loss and damage, island nations, African nations, and least developed countries offered a second ethical principle, or, rather, moral issue, to inform the process, one in which there would be an acknowledgment of the present and projected suffering coming out of climate change. The late entrenchment of loss and damage as a moral issue within climate negotiations and the ambiguous location of the countries that supported it put loss and damage in some tension with CBDR. However, developing countries tried to attend to this tension by assimilating loss and damage within their agendas.

Continuing with his story, Saleem said it was at COP18 in Doha in 2012 where the establishment of a mechanism for loss and damage became inevitable (Huq, Roberts, and Fenton 2013; McNamara 2014; van der Geest and Warner 2015). In a break from the usual practice, in which developed countries produced advance text that was shared and to which developing countries reacted, a new coalition of small island nations, least developed countries, African countries, and Latin American countries (by means of the AILAC bloc) came to the annual meeting with a prepared text on loss and damage, forcing the developed countries to respond.

The United States put the entire text in brackets, which made it optional or excisable from the decision text. Meanwhile the EU, putting on the friendly face of developed countries (as usual), shifted crucial paragraphs to the aegis of adaption and capacity building, signaling that while they were willing to be more cooperative, they did not want a separate track within the negotiations on loss and damage.

That year, Super Typhoon Haiyan devastated the Philippines, and Yeb Saño, the country negotiator for the Philippines (introduced earlier), gave an emotional speech at the opening session of the meeting about needing to face the havoc that climate change was already creating in people's lives. When the demand for a mechanism for loss and damage was deleted from the draft negotiation text, Juan Hoffmaister (an alumni of College of the Atlantic and former student of Doreen Stabinsky), who was a delegate of Bolivia and was leading G-77 and China on the issue of adaptation, got all the countries of G-77 and China to walk out of the negotiation room. Under pressure to come to an agreement, negotiations went into overtime until the developed countries "blinked," in Saleem's words (Vanhala and Hestbaek 2016). They eased up on their usual pugnacious approach to negotiations and accepted that their fellow, poorer nations needed an acknowledgment of loss and damage before negotiations could go any further. The 2012 COP decision text reincorporated the call for a mechanism. Finally, in Warsaw the following year (COP19 in 2013), the Warsaw International Mechanism for Loss and Damage (WIM) was adopted (Rajamani 2014). Loss and damage was here to stay.

Loss and Damage as a Clearinghouse for Information

Looking at what the WIM had been able to accomplish to date, one would be hard pressed to understand all the agitation and excitement around its establishment. As Saleem rightly anticipated in his blog post, WIM did no more than open doors to "research to minimize costs," "sharing knowledge," "insurance," "solidarity payments," and "risk retention." It served as a modest clearinghouse for information on best practices, rather than as a platform for testifying to the devastating effects of climate change and a means of channeling much-needed finance to those whose very existence was under the shadow of annihilation (Doelle 2014; Gewirtzman et al. 2018).

The mechanism was limited to examining best practices to provide the "poor countries" of Saleem's blog post with the different options they had with which to tackle loss and damage. But this knowledge was imparted as if the countries had deep reserves to draw upon to mobilize these options and as if these options would pay out sufficiently to cover the extent of devastation in the aftermath of disasters or over the long duration. WIM operated entirely in a hypothetical mode, and if "rich countries" didn't visit this house soon and more often, as Saleem urged, then even the wording of available options such as "insurance" or "solidarity payments" would quickly appear fantastical (Artis 2017).[5] Participants at WIM meetings complained bitterly that negotiators from developed countries did not attend them. And Christian charities took to berating developed countries for abandoning cherished values such as mutuality and solidarity, as activist demands for climate justice and equity also appeared not to take hold.[6]

Hoffmaister, a major force in pushing for the WIM, counseled patience in a personal letter to the public after the limited scope of the mechanism spelled out in the decision of COP19 was widely criticized. In 2019, Saleem assured me that it didn't matter that WIM was not a robust institution. The most important step was to have it down on paper. Procedurally international mechanisms had two arms—one technical and one financial. So far, the WIM had only been given a technical, knowledge-sharing mandate and was under constant threat of excision for having outlived this limited mandate. If, however, it could be made a permanent body under the Convention, then a financial arm was inevitable. And this arm would serve as the means to funnel funds to those in dire and urgent need. Saleem was sure that developed countries would come running to provide funds to the WIM or any of the other funds under the rubric of the UNFCCC rather than face the inevitability of the avalanche of legal action against them, one disaster at a time, one tragic story at a time, one case at a time. This was the long game.

With the WIM as a mechanism within the adaptation-focused Cancun Agreements, loss and damage was restricted to being an adaptation issue. After yet another hard-fought battle at COP21 in 2015, it was subsequently rescued from this silo by being made an independent article within the Paris Agreement (Mace and Verheyen 2016; see also Broberg and Romera 2020). The expectation for an article on loss and damage within the Paris Agreement beyond the WIM was that it would serve as a third pillar along with mitigation and adaptation. Its constant presence would spur the latter two spheres so as not to have to fall back upon the terms of loss and damage. And failing that, loss and damage would pay out where mitigation and/or adaptation fell short.

In its analysis of the adoption of loss and damage within the PA, TWN writes that some least developed countries were bought off at the last second by the United States, who, with their support, was able to insert a clause within the decision text of COP21 that undermined Article 8 of the Paris Agreement (Raman 2016). The clause read as follows: "Agrees that Article 8 of the Agreement does not involve or provide a basis for any liability or compensation." Those within the LDCs accused various countries of being the quisling.

Saleem informed me in 2019 that although developing negotiators fought the clause tooth and nail, ultimately its inclusion happened at the highest diplomatic level, advocated for by a prime minister of a small island nation. While disappointed with this qualification, since it seriously undermined the effort to give loss and damage teeth, Saleem was also philosophical about it. The clause was a "red line" for the United States, a non-negotiable item, and as such it would have to be contended with one way or another. It was better to get an article with a clause rather than not have an article at all. Loss and damage now had both a mechanism under the Convention and an article within the Paris Agreement.

At any rate, Saleem felt that legal action against nation-states was more the right of their citizens and actionable within national courts than international ones. And as a multilateral environmental agreement, the Paris Agreement was hardly enforceable in a punitive fashion, particularly not in its current mode of voluntary national contributions to emissions reduction and facilitative mode of encouraging compliance. What it could provide was "consensus" among Parties, which loss and damage had done. Everyone now accepted it as a legitimate term within the negotiations even if they did not agree on a shared understanding of what it was (Vanhala and Hestbaek 2016). Hence, Saleem felt secure in his vision that, one day, Article 8, the WIM, and other financial entities within the UNFCCC would be the logical ways to attend to loss and damage in the world (see also Mace and Verheyen 2016).

With most of the work on the Paris Rulebook concluded in Katowice at COP 24 in 2018 and with only the most contentious issues remaining, of which loss and damage was one, it was finally taken up in Madrid at COP25 in 2019. Previously I have spoken of how important host-country investment in the outcome of COPs is in guaranteeing their success, be it only in the form of a consensus decision text with no meaningful action items. When COP25 was shunted from Brazil to Chile but Chile backed out of hosting it because of the protests over austerity paralyzing the country, Spain clearly only stepped up at the last second to host as a courtesy to Chile. The convention was unfurled quickly and easily, as the hosts were used to an influx of tourists. It was held at a convention center at the edge of Madrid, easy to access by subway. Yet no one in the rest of the city even knew much of what was going on, busy as they were with Christmas festivities. And there was no decipherable political will or diplomacy driving the process such that one came away with a distinct sense of what investments the hosting country, which was Spain on behalf of Chile, had (Obergassel et al. 2020; Streck, von Unger, and Greiner 2020).

It ought not to be a surprise, then, that most discussions ended with the calling of Rule 16. This meant that all discussions were struck from record, and the consensus was to agree to act as if the discussions had not taken place at all. WIM was not granted a funding arm, as its advocates had hoped, but it endured to fight another day, and a decision was taken to create an expert group, the "Santiago Network," to catalyze technical support, encourage developed countries to scale up finance for loss and damage, and direct the Green Climate Fund to provide financial support (Allan and Yamineva 2019).

Without Hoffmaister managing the G-77 and China's support of the issue (he had moved on to the Green Climate Fund by this time), the shared understanding between developing countries faltered. This fracture showed up within the coalition of blocs that had been pushing for loss and damage since 2010. AOSIS and LDCs saw it as a governance issue, pushing for the WIM to remain under the auspices of the Cancun Agreements and, therefore, the Framework Convention, thus carving out loss and damage from the liability and compensation clause within the Paris Agreement. On the other hand, the AGN was interested in how it could be a means of accessing monies, such as a percentage of funds raised through public and private means, since finance through the expected routes had not been adequately addressed within the discussions on the Paris Rulebook.

Instead of seeing this as an internal quibble among Parties and blocs, I suggest that we see this as a conversation over the value of political rights versus material equality. Samuel Moyn's book *Not Enough: Human Rights in an Unequal World* (2019) provides us a point of entry for thinking of the discussion over loss and damage in these terms. Human rights are often considered as the highest achievement of humanity, in so far as they articulate ideals applicable to all. However, history shows that human rights emerged as a weak and compromised version of centuries-long discussions over rights, ranging from providing robust political rights to upholding material equality, with weaker and stronger versions of both. Moyn claims that human rights are the weakest version of what political rights could have been and that they completely fail to address the issues of social and economic justice.

I see the LDCs and AOSIS as concerned with strengthening the scope of political rights, with liability and compensation only an index of a robust suite of rights. Their preoccupation can be explained by the fact that, although not averse to loss and damage as a means of finance, they were concerned with unprecedented issues, such as the loss of sovereignty with the loss of territory and the mass migration and resettlement of populations. For the AGN, loss and damage wasn't only about securing finance for its own sake but about a means of wealth redistribution in a context in which African nations continue to be the site of violent resource appropriation and the political and economic instability it produces. Their contributions to the conversations about loss and damage emphasized the importance of material equality.

One might think of loss and damage as a flashpoint where important philosophical deliberations were taking place, even unconsciously and inadvertently, suggesting that human rights was never going to be the last word on these deliberations (Radin 1993; Calliari 2018; Verchick 2018; Calliari, Serdeczny, and Vanhala 2020). Just to provide another instance of this beyond the differences among the AOSIS, LDC, and AGN, one can compare the speeches of two world leaders in how they mobilized images of loss and damage toward two different ends, one in support of the free market and the other in indictment of it.

US president Barack Obama spoke thus at the opening ceremony of COP21 in Paris in 2015:

> This summer, I saw the effects of climate change firsthand in our northernmost state, Alaska, where the sea is already swallowing villages and eroding shorelines; where permafrost thaws and the tundra burns; where glaciers are melting at a pace unprecedented in modern times. And it was a preview of one possible future—a glimpse of our children's fate if the climate keeps changing faster than our efforts to address it. Submerged countries. Abandoned cities. Fields that no longer grow. Political disruptions that trigger new conflict, and even more floods of desperate peoples seeking the sanctuary of nations not their own.

Obama's discursive push was to present these dystopian images to emphasize that the time for action was now. Action for him meant a commitment to keeping temperature rise to 2°C within the century. This was an end to be achieved via alternative energy sources, with Obama underlining how much the United States had already done to contribute to this goal. Obama held out hope that the emerging market for alternative energy would direct us away from needing to be concerned with loss and damage. More generally, he was expressing faith in the free market to solve all.

In Obama's speech, images of catastrophe such as "Submerged countries. Abandoned cities. Fields that no longer grow" were presented in a hypothetical mode, as something that may or may not come to pass depending on how we act. However, these words already spoke to an entrenched and experienced reality. I quote extensively from Reverend Tufue Molu Lusama of the Tuvalu Christian Church, whose voice reverberated through the meeting halls in indicting the 2°C goal and the very economic system upon which Obama placed his hope.

In a letter addressing Parties to the agreement, Lusama (2017) wrote:

> Tuvalu is among the most vulnerable countries in the world to climate change and rises in sea level. We struggle to live from day to day. It is so hard to ignore the plight we are living in, and it is more difficult to explain to our people the reality and the link between environmental changes and what is happening on the ground. We continue to witness the loss of islands overnight, depletion of our corals due to bleaching, constant strong winds and cyclones, disappearing fish stocks, the dying out of our traditional plantations due to saltwater intrusion into our underground water tables and much more. As a Tuvaluan living in the

midst of these challenges, I see the injustices that cause our suffering. We have no part at all in creating this problem, and yet we are the first to suffer the consequences. How can the actions of developed countries—which create enormous amounts of CO_2 in the atmosphere and cause the global warming that leads to our weather pattern changes and rising sea levels—be justified? Why do we have to suffer because of something we did not do? The avariciousness and greed in the world economic system needs to stop. Justice needs to be carried out—for it is only through justice that countries in the world, like us, will survive in the face of this global challenge.

With the fate of loss and damage within the climate negotiation process, more specifically within the Paris Rulebook, as yet unsecured and likely to stay this way for the near future, let us now turn to the door of "liability and compensation." Saleem skips this door in his 2003 blog post, but it promises to help us understand the steep resistance to loss and damage.

As we saw earlier, the European Union was not opposed to loss and damage, providing it was assimilated within adaptation, much in the way that depreciation and inflation are added to goods to understand how the latter's value will fare in time. Also given that the EU was much more comfortable with regulatory structures and taxation, such as once provided by the Kyoto Protocol, they understood loss and damage as a form of tax to disincentivize pollution in the long run, similar to the Polluters Pay Principle advocated by organizations such as Bread for the World, Climate Action Network, and even Saleem (Hossain et al. 2021).

But not so the United States, which routinely chose to understand loss and damage as threatening liability and compensation. This fear made sense given that the United States had historically not been prone to self-regulation and that most existing regulations and protections within the country had been spurred by litigation. The United States probably correctly feared that the world would litigate against it, as it was among the richest countries in the world. The United States' resistance to loss and damage and the way the world understood this resistance didn't quite play out as I have put it. Gauging by some telling fieldwork encounters, I came to understand that the United States' resistance was understood as the country projecting its racialized politics onto the world stage.

Loss and Damage as Liability and Compensation

I first heard the term "loss and damage" from Saleem in 2012 at a climate change research conference at the University of Arizona, where I had gone to shadow him. He mentioned that he was just coming from a gathering of lawyers and other experts to discuss the possibility of pursuing loss and damage within the climate negotiations. Saleem called it "the crafting of the language on loss and damage." Likely he was referring to the advance text that would later be put forward at COP18 in Doha.

Even this early in our acquaintance, I had learned enough about his global reach as a climate diplomat to know that these words heralded a new addition to the global discourse on climate change, alongside mitigation and adaptation, which he was so crucial in shaping. I went back to my university and tried to discuss loss and damage with several of my colleagues who worked on environmental issues. They were nonplussed; it sounded to them like I was talking about reparations, and, based on their experience in the context of the United States, this conversation was a nonstarter. Having by that time lived in the United States for several decades, I knew reparations were associated with the political demands of the Black community for recompense for the free labor of their slave ancestors that was the historical foundation of the United States economy (Coates 2015).

In Paris in 2015, I made it a point of following loss and damage as a part of continuing to shadow Saleem within the negotiations. Activists and Party representatives of poor countries were regularly hosting side events and press conferences on the topic. One such press conference was held by CAN-USA. CAN, or the Climate Action Network (introduced earlier), was one of the two major networks that constituted the ENGO within the COP. They have been very involved in the negotiation process, putting out a daily newsletter, the *ECO*, which reported on the state of negotiations; organizing on different issues within the process to inflect their course; and holding press conferences and political actions in the conference space. CAN had been very invested in the issue of loss and damage, with Harjeet Singh of Action Aid International, Sven Harmaling of CARE International, and Julie-Anne Richards of CAN International serving as important mediators between disparate and oppositional Parties (Gach 2019). In fact, my interlocutors within DCJ would often complain that CAN imagined itself almost as a Party, getting deeply invested in minute details of the negotiations and losing sight of the bigger picture or broader political stakes of climate justice.

Similar to the regional representation of Parties within the UN, CAN too had representatives from every major area, including Africa, Europe, Asia, and the United States. At the press conference by CAN-US on December 2, 2015, I sat and listened as largely Black academics and activists from the United States spoke about the struggle to make reparations an important national issue within the United States and how this struggle gave them a vantage upon the issue of loss and damage being debated within negotiations at COP21. I remember being moved because I thought it was brave of US activists to travel to the COP to stress that the United States, a country that the world considered as so unified, instead was as internally divided and complicated as any of the other developed and developing countries.

I was at the press conference with Tanjir, a friend and representative of Action Aid Bangladesh. When he tore outside afterward, I rushed out with him. Before I could tell him how moved I was by the presentation we had just heard and how it surely gave ammunition to the issue of loss and damage, he hissed, "The Americans are at it again." He proceeded to tell me that the United States, never mind whether it was Party representatives or activists, had confounded the issue of loss and damage by bringing reparations to bear upon it. Reparations were the United States' internal issue. It was not an issue for countries such as Bangladesh and never would be. Bangladeshis could never make common cause with this issue, whereas loss and damage was about everyone. "We are all going to have loss that we can never recover from, and we need to have an acknowledgment of this fact. We need help because we can only do so much, not because we are owed compensation for past mistakes."

Fast forward to 2019. I was standing outside a hotel in Madrid, keeping two fellow Bangladeshis, a lawyer and a journalist, company as they smoked. They were discussing the deliberations on loss and damage underway in Madrid. The mandated review of the WIM was ongoing, as well as a consideration of how loss and damage could be addressed by making the WIM less of a clearinghouse for information and more of a mechanism by which tricky issues such as insurance premiums and rehabilitation could be addressed. In the midst of their discussion, the two stopped to bemoan Americans and their obsession with reparations. The journalist explained to the lawyer, "Initially I thought it was some internal issue, but I have come to realize that it is an existential issue for the Americans. The US [Party officials] sees us as their ex-slaves demanding justice, when really it's a matter of policy. What are we, all of us, going to do when people start to lose not just their homes but their countries? We have to have a plan, and all these people [Americans] are worried about how not to have to pay for their mistakes in the past. It is like a sickness. We say one thing, they hear something else."

Alondra Nelson in *The Social Life of DNA* (2016) recounts how genetic investigations begun in the 1990s in Buenos Aires and Johannesburg in forensic projects in support of human rights cases was by the 2000s put to the service of racial justice in the United States. Reparations for slavery were part of a larger attempt to address the question of racial inequality in the United States at a time when racism was considered to be "nonexistent" (109–10). Further on she writes that the idea of reparations, which draws on international law and discourses of human rights, was an important means by which to change the central figure of racial justice activism from that of "victim" to "creditor" (136).

It made all the sense in the world that activists for CAN-US would go to Paris to use the struggle for reparations as an object lesson for the discussion on loss and damage, to allow for a sharing of frameworks (international law and human rights law) and experiences (reparations and loss and damage), but it failed as a bridge. There was a disjuncture between loss and damage and reparations among its activists, even as the proscription of Article 8 within the COP21 decision text that prevented loss and damage as being the basis of liability and compensation had precisely the US concerns with reparations in mind.

Mulling on this disjuncture helped me to understand something that had long puzzled me about the discussion about loss and damage. When the AOSIS, LDC, and AGN spoke of loss and damage, I knew them to be talking about lost shorelines, salinity in the land and water, dead fish, and crumbling coral reefs, elements that accentuated nature as doomed. There was something downright confounding to me that these vulnerable countries had jettisoned the language of historical debt, putting aside any accounting for colonialism, genocide, resource extraction, and slavery, to advocate for money for climate devastation alone. I wondered: By what wizardry of accounting would all this rot and decay be linked to climate change and a monetary value put to it?[7]

Hearing my Bangladeshi friends complain, I realized that perhaps not unlike Nur Hashem, they wanted an acceptance of inevitable system failures and a plan with which to move forward. That was how Doreen Stabinsky, Juan Hoffmaister, and even Saleem would talk about it (see Bendell and Read 2021). So while it was fine to argue about whether the emphasis should be on political rights and/or on material equality, it was CBDR and reparations that dealt with the past, whereas loss and damage would be an issue of the here and now and the future. There was something deliberately foreshortened about this discourse as befitting an idea that could only be applicable to the world from 1992 onward. Thus, it should not be a surprise that Saleem invested considerable time and energy in helping the Ministry of Disaster Management and Relief in Bangladesh set up a national mechanism for loss and damage by which to demonstrate how such mechanisms would work in practice without belaboring the issue of who was to blame.

Loss and Damage Attribution

There is one last door for us to go through in Saleem's house of loss and damage. It is the one labeled "attribution," and it is one that keeps the door to "liability and compensation" shut. At the time that Saleem wrote his post, the science of attribution, that is, of attributing extreme weather events to climate change or to the greenhouse gas emissions of particular actors, was not as yet developed. It was only in 2014 that the IPCC made its clearest statement yet on anthropogenic climate change: "Human influence on the climate system is clear, and recent anthropogenic emissions of greenhouse gases are the highest in history. Recent climate changes have had widespread impacts on human and natural systems."

By 2018, there was already indication that attribution was strengthening. In Bonn during the May intersessional session of that year, I heard about the Peruvian farmer Saul Luciano Lliuya, who brought a case against the German utility provider RWE, charging that their emissions were causing glacier melt that was inundating his farmlands. The claim was rejected by RWE, and they would not entertain an out-of-court settlement, for fear that it would create a precedent. Although the lower courts in Essen, Germany, dismissed the case, the higher courts found the case was sufficiently well founded to allow for the gathering of further evidence (*DW* 2019).

In discussions on what has come to be called "loss and damage attribution," Christian Huggel, Daithi Stone, Maximilian Auffhammer, and Gerrit Hansen write in *Nature Climate Change* (2013) that we have to move away from thinking of climate change as a cause, as there is no one cause of weather events, but that climate change can contribute to weather events. When the authors turn away from science and look to insurance records on the incidence of weather events, it becomes clear that extreme impact events have increased in recent decades. The damage to assets provides a ready way to speak of loss. Moving away from cause-effect relations to thinking about contribution, such as through the likelihood of something occurring (statistical analysis) and the extent of its impact, provides a means to give definition to the vagaries of loss and damage. This is currently best captured by risk analysis, by which an event is broken down into its components of risk and each component examined for likelihood and extent.

Roda Verheyen, the lawyer fighting the case on behalf of Lliuya, has written that legal attribution ought not to be linked to scientific causation at all. Whereas the first relies not just on clear evidence but also argumentation based on a shared understanding of rules and norms, the second is necessarily qualified, being based on nonlinear processes, and is more often determined through falsification (Verheyen 2015). Falsification as a way to determine attribution for loss and damage poses a steep challenge, as these are unlikely to be lab-based experiments but more in the nature of observed and experienced real-world effects. We cannot possibly wait for these real-world experiments to run their course to be able to gather the scientific evidence needed to prove loss and damage. Hence, she argues for the separation of the two arguments, such that attribution can be argued legally, on the grounds of plausibility and sufficiency, rather than through a reliance on irrefutable scientific evidence alone (see also Hulme, O'Neill, and Dessai 2011; Huggel et al. 2013; James et al. 2019; Burger, Wentz, and Horton 2020; Lloyd and Shepherd 2021; Walker-Crawford 2021).[8]

The discussions that I am summarizing here are just the tip of the iceberg. It would seem that loss and damage attribution has begun to acquire more the quality of a forensic investigation, drawing upon multiple ways of examining a phenomenon to arrive at an adequate description of it. Huggel et al. (2013) write that while "science can assess how the occurrence of a particular event is related to anthropogenic greenhouse gas emissions, and how the damage is tied to drivers of risks," the aspect of blame that characterizes loss and damage attribution remains "fundamentally a political, societal or legal problem." They give "liability" as an example, pointing to how it is regulated differently across different national civil laws. As Saleem too remarked, ultimately the case for liability and compensation will have to be made country by country, case by case (Lusk 2017; Mace and Verheyen 2016; Marjanac and Patton 2018).

Loss and Damage as a Door to Mourning

At the first academic and activist conferences on climate change that I attended in the United States beginning in 2011, I was struck by how upbeat everyone was. One presenter after another got on stage and cheerfully explained climate models, the empirical data feeding into them, and the political contingencies and social dimensions of climate change the world over and discussed the future intensification of weather events, potential hotspots for struggles over limited resources, possible migration crises, fires, and so on, before going on to speak of programs and projects underway to help people make green spaces, get clean water, etc. I saw people in the audience shake their heads—I can't say what they were thinking, but I did not observe a single tear shed or angry sentiment voiced. That may have been why I was caught off guard by the climate scientist crying over the fate of Bangladesh at the conference in Oregon of which I spoke earlier.

It took me a while to realize that this projection of what was essentially a flat affect was necessary to make one's science remain separate from politics and potential charges of whipping up frenzy over climate catastrophes (see Danowski and de Castro 2016). It was crucial to making climate science take hold of the imagination, be convincing, without the suggestion of emotional blackmail or moral urging.

Given the widespread state of numbness in the United States in the face of climate change, it was with some interest that I noted how the language of loss and damage struck a chord with people. It was a matter of much contentious discussion at the COPs. It brought calculation much more forcefully to the front. So even if liability for loss and damage was not to be entertained as according to the text of the Paris Agreement, the fact that there was loss and damage meant calculability, if not yet causality. As such it provided a locus for dispersed emotions, giving focus to delegates and activists dissatisfied with having won the battle for insisting that mitigating climate change should be accompanied by adapting to climate change and now wanting to ensure resources for those poor and vulnerable already facing the effects of climate change.

I also found it to be the first instance within global climate change discourse in which there was a direct response to the fact of an incalculable loss. It brought darkness to discussions that were otherwise resolutely optimistic. The UN had always been committed to a gradualist picture of climate change and to the notion of scaling back its worst effects (Allan 2019). But for it to state the possibility of a loss that was unrecoverable opened an abyss within the climate change discussions yet to be reckoned with. Loss of tradition and heritage, loss of intrinsic values, stress, disorientation and other mental disorders, and pain and suffering, often ignored dimensions of climate change, have since found a place under the category of noneconomic loss and damage (Serdeczny, Waters, and Chan 2016). Loss and damage had become a place to park one's deepest fears and express one's darkest visions and most exacting experiences. Thus, in addition to the many doors through which loss and damage took us within Saleem's house, it also brought us to the threshold of mourning (Willox 2012; Das 2016).

Conclusion

The Gift of the Global South

I started this book by shadowing those within the scope of the Global South through the UN-led climate change negotiation process and end it with an analysis of the Paris Agreement negotiations and its implementation. In this Conclusion, I start with the world according to the Paris Agreement and work back to my initial question: What draws people to and keeps them within the COP negotiation process, as individuals, groups, or Parties?

The Paris Agreement was a voluntaristic agreement. What that means is that rather than laying down the objectives of what each country Party had to achieve in terms of carbon mitigation, it invited Parties to report their contributions to fighting climate change on a periodic basis. These nationally determined contributions were to show not only what Parties intended to achieve but also what Parties had achieved. For developing countries, especially for least developed and other categories of vulnerable countries, they could use these reports to project the inputs that they would need to achieve their intended goals. The Paris Agreement also embedded what it called a "ratchet mechanism," intended to increase the Parties' commitment to doing more over time. It invited Parties to update their NDCs without waiting for the preset deadline and to practice conscientious accounting to ensure the transparency and accountability of their climate actions. And in return, the Paris Agreement would mandate a periodic global stocktake of all Parties' NDCs and other related reports to see if the world was on track to keeping temperature rise within 2°C, and it would undertake facilitative dialogues with Parties, breaking the news to them as to how they were doing and advising them on how they could do more.

The entire enterprise is reminiscent of a progressive school for children whose parents cannot bear to discipline them and therefore instead smother them with passive aggression. It mostly suggests how jumpy and wary Parties were of any obligations upon them, be they developed or developing countries. And yet, there would be no way to tell where we were at in terms of progress without the agreement. As Richard Kinley, previously the deputy director of the UNFCCC, told me, with national pledges one just had public promises but no accounting of whether they had been met. With bilateral agreements, you had promises to each other but, again, no public accounting of whether they had been met. It was only in a multilateral agreement such as this that participants could keep tabs on one another. Thus far from designing, launching, and mandating broad-based climate action, the Paris Agreement was a mode of accounting for uncertain change, both climatic change and change in the behavior and outlook of nation-states.

This agreement was perfectly diagnostic of our present, with the parlous state of relations and deficit of trust among developed, developing, and least developed Parties. It was inflected by failures of historical reckoning, the lack of acknowledgment of wrongdoing, continued exploitation, and broken promises. If the developing countries charged the developed ones with these, they too were quickly falling into the same pattern of behavior with respect to their smaller, weaker, poorer compatriots, leaving the latter wide open to being "peeled off." This was a phrase that I heard with frequency, and I think it well describes the mode of manipulation at play. For their part, the developed countries preferred always to start in the present, even as the present moved swiftly into the past, so that it was only what prevailed at this very moment that was to set the conditions for their relations with most of the world.

And the Paris Agreement was what it was because the market mentality had come to successfully colonize, across large parts of the world, not just the global economy but also the sense of what was possible. The extent of this colonization was revealed by the fact that most UN officials and Party delegates were unmoved by claims that the process had been "captured by corporate forces," in the language of activists. Wasn't that precisely the point, to send a signal to the markets to give incentives to go green? To have private industry enter and be involved in the process and implicate themselves in the framework of accounting, transparency, and accountability? Adjudication seemed impossible, judging by acrimonious conversations between those who had utmost faith in the markets and those calling for degrowth.

All that the UNFCCC process was now committed to was ensuring that, once a common tabular format and timeframes for reporting had been finalized, everyone had the capacity to keep accounts and that they were doing so. It was a minimalist approach, easily criticized for abandoning all notions of actual, meaningful, on-the-ground interventions; reliant upon the goodwill of the changing governments of nation-states to follow through; and neglectful of safeguards against any marginalized communities within nation-states, despite the PA's grandiloquent Preamble. It took the ethics and sensibilities of the accountant and the auditor as the templates for meaningful political subjectivities and actions.

While the Global South continues to tug at my heartstrings and imagination, it was clear that there wasn't a well-defined concept of it within the process; instead, it was a placeholder (not unlike the Like-Minded Developing Countries, or LMDC) occupied by different groups of countries at different points. "Global South" designated a minimalist approach to South-South solidarity. I was told that major developing countries were in many different bilateral relations with the smaller, more vulnerable ones but that these were kept out of view of the process, to keep pressure on developed countries to step up.

So then why were people engaging in this process? Was it just force of habit produced by negotiations that had been going on for thirty years? I now had a better understanding of what it meant to say that one had to engage it because it was the only global process. This explanation wasn't just about the scale of the process but rather about the fact that it was the only process by which Parties were bound to share some basic protocols of self-reporting, which made them "transparent" to one another, although one could easily cast doubts on whether numbers and metrics were quite that transparent or even hooked to reality.

But we know that the fact it was the *only* process wasn't the *only* reason people participated in it. People came as parts of Party delegations, political blocs, and the Global South. Even besides that, people came of necessity; it was a job, and they had to do it. Some were motivated by the thrill of making deals with fellow Party delegates or making a business score. Some came to see it as a calling. They came because they wanted to do something about climate change, and doing this felt as good as doing other things. Some used it as the opportunity to gather to create and strengthen global solidarity movements. Some saw in it the possibility of a future world order governed by rules and procedures. Some came to serve as monitors, to ensure that their Parties knew that they were being watched. Some came to change the discourse. An interlocutor within the UNFCCC Secretariat pointed to the scores of people streaming past us to emphasize that the process had created them as much as it created the Agreement. Without the process, these people wouldn't be here doing whatever they were doing, which at the very least was keeping the attention on climate change alive.

COP was like a gigantic wasteful rave for the climate, but, like many rituals (and perhaps like any successful ritual), it was productive of an excitation that then effervesced out into the wider society. As an anthropologist, for this reason alone I am loath to write off the process. Anyone who participated in the People's Forum within the heart of COP26 in Glasgow will understand what I mean.

Bangladesh in Madrid and Glasgow

Bangladesh had a bumper presence in Madrid in 2019 and then in Glasgow in 2021. Unlike previous years, when its delegates had to be satisfied with a booth, this year the decision was made to pay for a country pavilion. Zia bhai explained that the government of Bangladesh felt that there were enough Bangladeshis in the process to justify the expense, plus it helped put Bangladesh in the public eye, plying its Janus-faced approach as both the future of climate catastrophe and climate opportunity.

The pavilion was the center of quite a lot of gathering and activity. Just sitting there at various times, I saw the main seminar room fully occupied with side events and the main office and conference room fill and empty throughout the long days. The front desk was manned interchangeably by junior government officials and young people associated with ICCAAD, the research institute Saleem ran in Bangladesh. Reporters, civil society representatives, interns, and the odd researcher such as myself milled about, or lounged on the comfy seats around the front desk, or stood in groups in front of the pavilion, shooting the breeze or making plans. It felt like a party.

Zia and Shawkat were not too busy at the Madrid or Glasgow COPs. Decision making relating to mitigation and adaptation had been wrapped up in Katowice. Hafij and Mizan, the two civil society members of the official Bangladesh delegation who looked at loss and damage and finance, respectively, were much busier. While Mizan was mostly reporting, Hafij was more involved in the nitty-gritty of negotiation and had little time to spare for conversation. While my interlocutors in other parts of the negotiations, such as activists, those involved with other developing country Parties, and even a few officials, were despondent about how the negotiations were going, the mood among the Bangladeshis was upbeat. Once again Saleem provided insight. He said that if one didn't get too caught up in all the technical details and took a more zoomed-out perspective on contentious issues, it was clear that whether finances were coming or not, loss and damage was now part of the future of these negotiations.

Youth Action

The importance of involving youth in climate-related issues was there from the start of the process, at the 1992 Rio Summit, with youth (defined as between the ages of fifteen and twenty-four) considered important partners in sustainable development. However, as we learned from Sebastien, they were late to organize within the UNFCCC because they were uncertain whether they wanted to be part of the formal process. Perhaps they might be more effective and less bound by rules if they organized informally. However, once instituted as the YOUNGO, they were very effective, running a full roster of events during the day earmarked for youth at the COP sessions and intersessions, lobbying delegates of their countries on relevant issues, and organizing with other CSOs to carry out public actions within and outside the sessions. Perhaps the strongest youth showing was at Glasgow, whose numbers of participants far exceeded even COP21 in Paris.

In *"Youth Is Not a Political Position": Exploring Justice Claims-Making in the UN Climate Change Negotiations* (2020), the authors Harriet Thew, Lucie Middlemiss, and Jouni Paavola provide a rare view into the evolution of the youth position within the climate negotiations. Exploring the nature of the justice claims made by a youth group based in the United Kingdom, they show how the youth constituency went from protesting climate injustice in terms of risks to future generations to expressing solidarity with those experiencing injustices in the present. While Greta Thunberg, the Swedish youth activist, was widely seen as increasing youth participation, several members of the YOUNGO from the Global North, whom I interviewed, recounted the dismay they felt when they first encountered members of civil society from the Global South and realized the extent to which their youth agenda was steeped in privilege. This encouraged them to branch out to be more inclusive of a broader range of issues, leading Tetet to speak approvingly of hearing southern arguments issuing from northern activists.

This broad-based coalition was visible within the Preamble to the Paris Agreement and continued to be visible at all the sessions I attended. The most powerful action I was involved in with youth and civil society organizations was telling both of the inclusive message coming from youth activists but also their increasing marginalization within the wider climate negotiation process. After Greta decried "shame on you" to the delegates for their obvious lackluster progress in Madrid in 2019, two hundred or so activists decided to stage a spontaneous rally and march in the midst of the meeting halls in Madrid, without prior permission from the Secretariat. The decision was unusual because the Secretariat tended to be accommodating of public actions, although usually cordoning them off in specific areas sometimes far from the actual negotiations. The idea was to shake up the process so that negotiators would face public anger and upset head on, rather than attend a spectacle advertised over CCTV, photographed, and quickly put aside.

We were told by the organizers to listen for a whistle, then quickly gather in a central space with metal cups, bottles, and spoons so we could make a beat to accompany our chants. Among the chants circulated were "Stop the corporate profits. End all carbon markets" and my personal favorite, "We rise before the seas. We rise as tall as trees. We rise not just for you and me. But for all humanity. We Rise for Liberation. We Rise for all Creation."

No sooner had we assembled than security personnel crowded around us, corralling us a little way down the hallway and through an enormous doorway leading to what looked to be an outdoor, enclosed loading area. They marched us outside and closed the doors behind us. We stood there rallying some more. I turned around at the sound of a familiar voice and found myself face to face with no one other than Asad. He looked startled, but after he had placed me, he beamed and said, "Good to see you here. Good for you," and, with those friendly words, dismissed me and returned to his conversation. I had a feeling of déjà vu.[1]

It was unclear what was to happen to us, whether we would be barred from attending the rest of the COP in Madrid and even future COPs as well. A few of the civil society organizations liaisons, Nathan among them, went and argued our case with the Secretariat, which then agreed merely to deprive us entry into the session for the rest of the day. We were let out of the loading dock, our passes taken as we exited. This step, construed as a punishment, meant that we needed to get new IDs the next day and was clearly directed at letting activists know that nothing they did could deter negotiations—nor was their presence central to negotiations.

While Thew, Middlemiss, and Paavola appreciate the efforts made by youth from the Global North to pursue solidarity actions, they also make clear that these actions impeded the mandate of the youth to represent future generations. They note that "youth" is no longer seen as a political position, only as a quantity of bodies thronging in support of the actions of others.

The Gift from the Global South

I want to end with a final reconnoiter of the process, not from the perspective of what has been, what is, or what is to come, but that of *what if.* People at all levels of the climate negotiation process and elsewhere form relations for a variety of reasons, from group identification; to diplomatic, strategic, or pragmatic alliances; to solidarity expressions, coalitional politics, and relations of care. The relations are informed by political or economic interests and inflected by standing relations. They are infused with trust or distrust, among many other feelings.

What is missing and constantly remarked on is an equitable relationship. One might say that the nature of power is to constantly create imbalances, ensuring flow through the body politic and the international order. The existing state of international inequality and inequitable relations are very much the product of history and contingency. Within the anthropological archives there is one relation that has been studied that does not so much redress inequity or power imbalance as make it into a positive force, a force for creating community. This is how the *gift relation* is understood (see Mauss 2002; Hénaff 2019). The giving of a gift produces a strange kind of imbalance. The gift, given with no demand, nonetheless embeds reciprocity into the relation as something due sometime in the future.

Throughout my research, I kept mulling over what made Bangladesh anything special or anything at all to study, beyond the fact of the accident of my birth there and continuing link through field research. It is just one more run-of-the-mill poor country. Yet, as with the study of any particularity, it yields important perspectives on the state of being bound, of being obliged, and how one relates to it. Unquestionably, Bangladesh is bound, but it occupies that state in interesting ways. Entangled with developing countries through regional treaties and trade, it expresses ambivalence toward them. Dependent on developed countries for aid and trade, it espouses pride in retaining its sovereignty and having an up-to-date market ethos.

Bangladesh did something a bit out of the ordinary in pursuing the issue of loss and damage in the face of its relations to other developing countries and to the developed world, for which it cannot really provide a good justification. Pursuing loss and damage endears Bangladesh to neither developing nor developed countries. Most involved with the issue in Bangladesh think of it as a possible opening into the future, that something, as yet inchoate, might just come out of it. What if Bangladesh was to give the issue of loss and damage as a gift to global youth, in the high ritual fashion that marks the occasions of gift giving within the anthropological literature? The ritual isn't a performance or a spectacle. It is to indicate the seriousness and sincerity of the gesture. And I like the idea of a country whose economic dependence is clear for all to see and whose political credentials aren't quite as transparent and pure as, say, Venezuela's or Bolivia's (at least with regard to COP) as the one giving the gift. What if this gift of the tools provided by loss and damage allowed the youth to regain their mandate to represent the future and to fight for it? And as is the nature of gifts, who knows what will be reciprocated, or when, but something, at some point, will be returned. The anthropological literature shows that it always is, for better or for worse.

Acknowledgments

Many people have been very generous with their time in making this process come to life for me. My biggest thanks go to Dr. Saleemul Huq for inviting me to join him at the COPs. He and his team of young researchers at ICCCAD have been warm and welcoming on every occasion, and I have learned more from him than I can say. The fact that he is in almost every chapter speaks volumes of not just his knowledge and expertise but also his dedication to this process and his commitment to empowering people through encouraging them to learn. I am also very grateful to Saleem for introducing me to the work of the International Institute of Environment and Development (IIED) and in particular to Clare Shakya and Achala C. Abeysinghe. I have learned a lot from IIED's Climate and Development Days and about the negotiation process and nonstate actors' participation in it from these admirable women.

Ziaul Haque and Shawkat Mirza of the Bangladesh delegation were similarly warm and embracing, and I received an education from them ranging from ministerial interrelations to the intricacies of climate diplomacy. Hafij Khan, the environmental lawyer who was among the nongovernmental negotiators in the Bangladesh delegation, and his partner, Sharaban Tahura Zaman, in the delegation for Gambia, made me part of their family. I wish for these "young 'uns" every success because they certainly do Bangladesh proud. I am also very grateful to have met and spoken at length with Dr. Nurul Quadir, of the official arm of the Bangladesh delegation, and Dr. Mizan Khan, who followed finance for the Bangladesh delegation. I always met Farah Kabir and Tanjir Hossain, of Action Aid; Mollah Amzad Hossain, reporter extraordinaire for *Energy and Power Magazine*; Masroora Haque, then of IIED; and Jahangir Hasan Masum of Coastal Bangladesh at the COPs with great excitement for the pleasure of their company as much as for the insights they had to impart to me.

I thank Asad Rehman, previously of Friends of the Earth and more recently of the World without Want, for his patience in having me pop up all the time while he was busy organizing and allowing me to come along for the ride. The lead organizer of Global Campaign for Demand Climate Justice Nathan Thanki's skepticism toward yet vast knowledge of the process was very grounding and much appreciated. Lidy Nacpil, of the Asian People's Movement on Debt and Development, was kind enough to allow me to shadow her. It was inspiring just to be in her presence. I tried to hang around Matthew Stilwell, of the Institute for Governance & Sustainable Development, and Mark Jariabka, of Islands First, as much as I could because although they were involved in the process in diverse capacities, they also kept up a penetrating structural analysis of the goings-on around them, making me realize the extent to which intellectual analysis and self-reflexivity could accompany the process. Tetet Lauron, of the Rosa-Luxemburg-Stiftung, was just a pleasure to be around, both for her infectious joy of life and refusal to be browbeaten by authority. She goaded me the most in terms of my lack of involvement in activism, pushing me to be more involved and to care more, not abstractly, but in the here and now. Sébastien Duyck, of the Center for International Law, was very patient in spelling out the significance of human rights for the climate discussions and the climate issue for human rights.

It was an honor to meet Richard Kinley, the now retired deputy executive secretary of the UNFCCC. Through conversations with him I came to better realize how the United Nations once held out the promise of not just an exciting international career but a calling to make the world better through diplomacy and multilateral engagement. Dr. Youssef Nassef, of the UNFCCC Secretariat, was a visionary hiding out in plain view. Lindsay Cooke, of the Quaker United Nations Office, along with Tetet, made me realize the importance of tapping into the dimensions of the emotional and the spiritual in the frenzy of the negotiations.

Rochelle Tobias and Debbie Poole, my colleagues at Hopkins, were my fellow adventurers. Even though neither works on climate change, they came with me to Paris and lent me their eyes and shared their experiences. I am not sure I would have had the courage to jump into this labyrinth to do research were they not such willing early research partners. Ben Zaitchik, of the Earth and Planetary Sciences Department, spent lunch hours drawing on his years as part of the US delegation to explain the process to me before my first COP. I spent a fun evening with Ben and Judy Orlove in Paris, and Kasia Paprocki and Mariam Banahi dropped by at various points in Bonn, for whose terrific cheer I was very grateful. Later I came to know Julie Raymond, Magnús Örn Sigurdsson, Jessica O'Reilly, and Cindy Isenhour, fellow anthropologists, with whom I have spent many evenings eating marvelous meals while trying to connect the dots of this process. Julie was an incredible organizer, getting us passes and places to stay. Magnus generously made important introductions that helped me in terms of access and gave my first full draft of this book a serious read. I hope our nourishing conversations continue for a long time. Danielle Falzon, of Brown University, whom I met as often in Dhaka as at the COPs, was a tremendous source of scholarly citations when I was stuck on the accounting and accountability issue within the Paris Agreement.

I somehow managed to give a talk in person on a chapter of this book at the Program in History of Science and Medicine at Yale University just before we all retreated to our homes because of the pandemic. That was a wonderful occasion of conversation and exchange facilitated by Deborah Coen. I thank also the Chowdhury Center for Bangladesh Studies at the University of California–Berkeley, in particular Sanchita Banejree Saxena, for providing the virtual occasion to present some of my work; Kasia Paprocki and Jason Cons, my co-panelists; and Sugata Ray and Sarah Vaughn, our fabulous discussants. Ken Alders and Sarah Carson, in the Science in Human Culture Program at Northwestern University, and Taushif Kara, of the Center of Islamic Studies at the University of Cambridge, also gave me the opportunity to present my work virtually, which allowed for very welcome feedback during the long two years of self-isolation. Sophie Haines kindly put together an event at the University of Edinburgh to allow Magnús Örn Sigurdsson; Perry Maddox, a graduate student at Johns Hopkins; and me to present on our research on the COPs. I also note with admiration James J. A. Blair and Cindy Isenhour's herculean efforts in bringing together in conversation anthropologists working on the climate negotiations in what promises to be a terrific collection for *Cultural Anthropology*'s Hot Spots.

I also want to take this opportunity to thank the students in my class on "Climate Change and Everyday Life" (co-taught with Rochelle Tobias) and "Climate Change: Treaties and Politics," for being game for hearing my preliminary thoughts and reading my manuscript as it emerged in dribs and drabs. My graduate students, in particular Burge Abiral, Perry Maddox, and Sumin Myung, deserve thanks for reading and giving very useful input on a draft of the book. My friends Sharika Thiranagama, Andrew Brandel, and Swayam Bagaria read all manner of things on which I write and on which they have no interest, but they somehow always manage to give sage and trenchant advice. They extended the same care to drafts of this book. As always, I am in debt to them.

Johns Hopkins University has been a great home for this project. My colleagues have always been patient in hearing what I have had to tell them about the negotiations after I return from the COPs bursting with things to say. I received a much-appreciated Catalyst Award, which helped subsidize my research. Most recently I was awarded the TOME Monograph Subvention grant to allow this book to be open access.

I thank Tom Lay, my editor at Fordham University Press, for seeing something in the book that made him take it up quickly to have it reviewed. I appreciate the faith he shows in my work, and I hope it lives up to his expectations. As always, I thank my loving family, Bob, Sophie, and Suli, for being so patient with my absences and the pressures my research places on them.

I didn't really set out to study this process. I set out to study climate change in Bangladesh in its many shapes and forms and serendipitously followed an interlocuter to COP21. This detour turned out to be momentous, as that was where the Paris Agreement was signed into existence. I have spent the last eight years trying to figure out the importance of that moment in human history, which I caught quite by accident. This book is a small contribution to considering the significance of the Paris Agreement, specifically for countries like Bangladesh. It is more retrospective and presentist than future oriented, whereas a proper judgment of that historical moment awaits the future, as the Paris Agreement is only just gearing into action. To understand the Paris Agreement meant immersion in the UNFCCC negotiation process, which I undertook mostly to solve the puzzle of how it works, but over time I have come to feel that my disinterring of the process, or at least some parts of it, could be useful to others. I hope that those working on the issue of climate change in myriad ways and wanting to understand how this global process has any bearing on their lives or where their actions sit alongside it may find it illuminating. This exercise of thinking one's degrees of separation from the process could engender excitement for how to get involved and make a difference from within or outside the process, or it could well produce frustration at how incremental and unmoored from reality everything feels about it. But at least people would know the limits and possibilities of the process without being incautiously enthusiastic or dismissive of it.

This is what I have to offer. I offer it thinking of the children that my family, myself included, have brought into this world and consigned to an uncertain future. I hope that this book may serve as a useful guide to a small part of that future. I dedicate the book to my children and to my nieces and nephews.

Notes

Introduction: The Climate Regime

1. In the strong versions of this argument, developing countries are owed an ecological debt proportionate to the extent of the devastation wrought by industrialization by developed countries (see Goeminne and Paredis 2010).

2. There are countless writings assessing the efficacies and possible failures of the Paris Agreement, which I assess in the chapters ahead in lieu of adjudicating this in the Introduction.

1. How to COP

1. This chapter takes its title from the UNFCCC publication: *How to COP: A Handbook for Hosting Climate Change Conferences*, https://unfccc.int/documents /185998.

2. Friends of the Earth International is considered a highly effective environmental organization because of their bottom-up mode of affiliating and creating global solidarity movements. See, for instance, Duncan McLaren (1992) and Brian Doherty and Timothy Doyle (2013).

3. Both Mattias Hjerpe and Björn-Ola Linnér (2010) and Daniel Klein et al. (2017) discuss how side events were a very important means by which civil society participants were educated into the shared conceptual basis of the negotiations and provided input into the negotiations. While I came to share this understanding, I think that these scholars may go too far when they claim that side events helped generate a shared vision among the participants, because there was no clear site where a shared vision was demonstrated either within these events or within the official negotiation space, which were marked by sharp divisions, as I explore in what follows.

4. The Green Climate Fund's primary activity was to provide funding to developing countries to assist with mitigation and adaptation goals. See S. Niggol Seo (2019) for a brief history and overview of this entity. It was one of several financial mechanisms within the climate regime, others being the Global Environmental Facility, Special Climate Change Fund, Least Developed Countries Fund and Adaptation Fund. For an overview of these entities, see UNFCCC, "Introduction to Climate Finance," https://unfccc.int/topics/climate-finance/the-big-picture/introduction-to-climate -finance.

5. Rikard Warlenius (2018) shows how the primary claim of climate justice was to have industrialized nations repay a "climate debt" owed to developing countries, which constituted "returning," and thereby "decolonizing," their environmental space damaged by unjust economic exploitation. Furthermore, he distinguishes between "ecological debt" and "climate debt," in that ecological debt asked for the cancellation of external debts on grounds of such exploitation, whereas "climate debt" asked for the restoration of communal rights and territories. See also Jekwu Ikeme (2003), Bradley C. Parks and J. Timmons Roberts (2010), Antonin Pottier et al. (2017), Evan Gach (2019), and Md. Fahad Hossain et al. (2021) for further discussions on the notion of justice within the climate negotiations.

6. Friman and Linnér (2008) show how historical responsibility represents the effort to correctly represent physical nature, specifically emissions of the past, in climate models (see also Matthews 2016). However, they claim such efforts miss the original intent, which is to make equity the guiding principle in North-South relations. This is what I see Asad and other activists as doing.

7. The list of demands that the DCJ put together at the conclusion of the two-day assembly in Bonn had a rough and tumble feel to it, resistant to easy translation into negotiating points. DCJ members were insistent that they be taken in their own terms in order to produce what Asad referred to as "a change in discourse." Of course, such a change also ran the risk of activist language being co-opted. For instance, at COP23 I found the word "justice" so widely deployed as to be empty of content. It very seldom referred to historical responsibility, yet the two were linked for DCJ, which had been foundational in bringing the concept of justice within negotiation discourse (Hadden 2015).

8. Hadden (2015) shows this to be the case for the activists who attended COP15 in Copenhagen, which is often referenced as a failure to be avoided at all costs within the process. However, a closer examination of Copenhagen reveals it to be a productive failure in so far as so many consequential actions stem from it. The conference was badly managed by the Danish government, which had prepared a decision text ahead of the meetings that was only shared with a select group of the Parties to the Convention, notably those among the developed countries. Once word of this shortchanging of the negotiations got around, it eroded both trust in the Danish Presidency of the COP and relations between developing and developed countries. This was the first COP in which there was no unanimous decision regarding action but rather an accord to continue to converse (Marsden 2011). This was also the COP at

which environmental activists broke away from the accommodationist politics of CAN to undertake more confrontational politics demanding meaningful climate policy with justice considerations. It ended in violence, police retaliation, and the conviction of Danish activists. This confrontation in turn led to a falling out between civil society members and the UNFCCC Secretariat and to the implementation of more stringent rules on who could participate in the COPs and under what conditions. According to DCJ, this COP provided a stinging lesson for those in CAN that they couldn't simply accommodate themselves to the negotiation process but had to bring issues of concern of their local constituencies to the process. It was vital for developed countries because they got buy-in from the major developing countries that Kyoto was not the way forward but rather that there was need for a new agreement that would include developing countries pulling their weight in mitigation efforts. For this concession, developing countries received the promise of fast-tracked finance of $30 billion a year, which would be ratcheted to $100 billion a year as baseline from 2020 onward (see Haug and Berkhout 2010; it was this promise to which Asad was referring). According to Saleem, COP15 was a learning experience for developing countries and was important in shaping their determination to invest their energies in loss and damage in subsequent COPs.

9. Fair share, as I came to realize, belonged together with the concepts of climate debt and historical responsibility discussed earlier and together comprise the elements of the argument for common but differentiated responsibilities (CBDR), the principle of international law that undergirds the climate negotiations by which all are held responsible for addressing environmental damage but not equally responsible. https://www.britannica.com/topic/common-but-differentiated-responsibilities. See Tuula Honkonen (2009) for a history and regulatory framework for implementing CBDR in environmental agreements. In the schematic history that Honkonen provides, we see how CBDR takes off only after the failure of resource redistribution attempted by the Global South through the New International Economic Order in the 1970s (see also Getachew 2019). CBDR may be seen as a last-ditch effort to retain equity as an organizing principle within international law, even if just restricted to environmental issues.

10. See https://www.wearestillin.com/about.

11. Historically, this part of Poland had been a part of Germany, and its mineral resources had served Germany during the First World War and to help pay for postwar reparations. Poland had long laid claims upon the region, and its own inhabitants had risen up several times against German domination, in what came to be called the Silesian Uprisings. Poland was only able to reclaim a part of it with the help of the Soviets in the mid-1940s.

2. The Voice of Bangladesh

1. See Kasia Paprocki (2021) for an account of Nigera Kori's history and courageous activities on behalf of the landless poor in Bangladesh.

2. This aspect of side events is not adequately captured by Hjerpe and Linnér (2010), who focus more on how side events entrain participants into the conceptual framework of the climate negotiations, test out new ideas at all scales, and create a shared vision for climate.

3. The Green Climate Fund was the financial mechanism created under the UNFCCC to provide funds to developing countries to assist them in tacking mitigation and adaptation practices. It is one among several financial mechanisms serving the climate regime. See Seo (2019).

4. See https://unfccc.int/process-and-meetings/Parties-non-Party-stakeholders /Parties/Party-groupings.

5. The one very interesting development in recent years was the effort by countries in the AGN to enter into the "special circumstances" category historically accorded LDCs and SIDS, which gives the latter flexibility in terms of mitigation goals and greater claims upon support in building up their capacity to combat climate change (Chan 2021a). Their efforts for inclusion had been repeatedly blocked by certain countries. As policy was historically only made by means of consensus within the process, even the resistance of a small number of countries was sufficient to block such efforts. A closer look at the situation reveals that it was countries in Latin America who did not meet the criteria for special circumstances but who also did not want to grant more developing countries this protection who were blocking this petition, thus further complicating the picture of solidarity among developing countries that G-77 and China was attempting to project.

6. After this phrase "peel off" was used in several conversations I had with various activists, I decided to look up its meaning. I had taken it to refer to developed countries or large developing countries inducing small and/or weaker countries to act otherwise than they would normally. The idioms portion of the Free Dictionary said that it meant "to come off of someone or something in thin stripes or pieces" (https://idioms.thefreedictionary.com/peel+off+from+someone), while the Urban Dictionary of slang words indicated that it meant "to run away from something, very quickly, either in escape of violence, or police" (https://www.urbandictionary .com/define.php?term=Peel%20Off). Given that the word was used colloquially, I imagine all these idiomatic and slang uses were in play.

7. See Lau Øfjord Blaxekjær et al. (2020) for how LMDC mounted arguments against such charges.

3. Who Wants to Be a Negotiator?

1. See Matthew C. Nisbet (2009) for a discussion of how important frames are for the perception of climate change. It was clear from Abdulla's words that he was attentive to this aspect of climate change within the negotiations.

2. See Carbon Brief (2021) for such lists relating to the COPs. Brazil had been known to send up to five hundred people as part of its delegation.

3. Saleem could also reasonably take credit for broaching the possibility of the CVF with President Nasheed of Maldives before COP15 in Copenhagen.

4. ecbi and IIED were among a handful of organizations providing such training to negotiators from the Global South. Also among them was CDKN (Climate and Development Knowledge Network; see Jefford and Hamza-Goodacre 2013). Various organizations, including IIED, also provided easy-to-use pocket-sized books on how to negotiate, with titles such as *Becoming a UNFCCC Delegate: What You Need to Know* (https://pubs.iied.org/sites/default/files/pdfs/migrate/17385IIED.pdf) or *Climate Negotiations Terminology: The Pocket Guide* (https://pubs.iied.org/sites/default/files /pdfs/migrate/10148IIED.pdf).

5. I immediately took a tally of the kinds of meetings I had been able to attend to date. While this list has grown, here I include only those I attended at the time of writing this book. They were the COP opening and closing plenaries at several COPs, numerous contact group informals (PA Article 4, APA agenda item 3), an informal informal (Bangladeshi negotiators meeting with the chairs of the contact group for APA agenda item 3), a corridor discussion (of the LMDC, upon which I shamelessly eavesdropped), and a huddle (of G-77 and China within the contact group meeting on the fate of Consultative Group of Experts [CGE], which the United States and Saudi Arabia began trying to excise in 2017, with pushback from LDCs). In addition, I also attended an informal bilateral meeting, one between Bangladesh and the EU, which was beyond the UNFCCC's scope of transparency and involved Parties meeting one another on specific issues. The one I attended involved the EU "taking the pulse" of Bangladesh on the issue of LMDC, to see if they were considered blockers or not by Bangladesh.

6. I was to be reminded of Fry's depiction of negotiations when I met and chatted with a delegate in Glasgow in 2021.

7. Robert Falkner (2016) suggests that the Paris Agreement, with its emphasis on voluntary participation, sought to capitalize on this mode of nudging recalcitrant Parties into action.

8. See Katharina Rietig (2014) for an overview of the different types of experts and the inputs that they provide within the climate negotiations. See also Chan (2021b) for case studies of NGOs providing expert guidance to Party delegations within the process.

9. See https://climateanalytics.org/.

4. Politics in Between-Spaces

1. We met in person at COP26 in Glasgow, at which he had come out of retirement to help the UK government with setting up the session. It was a short but invigorating talk in which he walked me through the new buzzwords at the session, the exciting developments, and the inevitable dead ends to be anticipated at Glasgow (also see Depledge, Saldivia, and Peñasco 2021). He made one very important distinction between country-based and multilateral actions that suggested to me why the Paris

Agreement is the agreement for a world in the state that it is right now with respect to international relations. When I asked him why the leaders of the world came earlier than the second week to COP26, the practice to which I was most accustomed, he told me that COPs can take both forms, with leaders coming either in the first or the second week. He said that they came in the first week when the emphasis was on encouraging ambitious action without expecting any, whereas they came in the second week when they could be sure of having something to congratulate one another on. All the countries that came or participated remotely (notably China and Russia) made national pledges of various kinds, which added up to impressive climate action. However, Kinley pointed out, the fact that they were national pledges placed them outside of the scope of the Paris Agreement and therefore unable to be monitored unless they showed up in countries' NDCs. In other words, as outside of the scope of monitoring enabled by the Paris Agreement, they were entirely performative.

2. Available at https://www.youtube.com/watch?v=8Rz8zV1f-S4.

3. Profiles of Paris, https://profilesofparis.com/.

4. In Glasgow in 2021, Kinley expressed great dismay that the developed countries in the world had not yet come up with the $100 billion dollars per annum by 2022 earlier promised to developing countries because, as he pointed out, if the COVID-19 global pandemic had shown anything, it had shown how much money governments of the world could mobilize at short notice to attend to an emergency. In other words, the countries of the world still did not think of climate as an emergency, nor were they willing to aid developing countries. In our interview, he also said that he feared that the Global North had decided not to budge on finance until such time as China declared itself a developed country and removed itself from the G-77 grouping, to the relief not just of developed countries but also smaller, poorer developing countries.

5. See https://unfccc.int/files/Parties_and_observers/ngo/application/pdf /constituencies_and_you.pdf.

6. See Cinnamon P. Carlarne (2010) for an interesting history of the rise of an environmental religious sensibility within the context of the United States that has spawned many Christian efforts at advocating for a more progressive agenda for tackling climate change.

7. See https://www.arrcc.org.au/living_the_change_faithful_choices_for_a _flourishing_world.

8. See https://quno.org/timeline/2016/1/video-lindsey-fielder-cook-talks-about-our -human-impacts-climate-change-programme.

9. Tetet told me of a time in which she saw Hafij, also a friend and the Bangladesh negotiator on loss and damage, be the subject of collective bullying by developed countries within a negotiation meeting. It was so humiliating that civil society constituencies sitting in the darkened portion of the hall shrank, not knowing how to respond. I imagine within such a mode of politics, having a safe space to speak with those who bully you during the day could be transformational.

10. The 2018 Talanoa Dialogue was how the Fijian Presidency of COP23 carried out its mandate to facilitate a discussion among the Parties as to whether they had been sufficiently ambitious in their first round of nationally determined contributions in reducing greenhouse gas emissions and help them move toward greater ambition. The structure of the dialogue aimed at fostering the sharing of stories by representatives of each country and members of civil society as to how they were faring under current conditions, what they were doing to combat climate change, and what they were striving toward as the means to enhance ambition. I participated in the dialogue by speaking to the issue of the current state of life in the island *chars* where I carried out my fieldwork in Bangladesh and how they were likely to be affected by climate change.

11. While economic anthropologists have always studied economics from the human perspective, more recently there has been an upswing of such writings that ask how the economy can become more human, how people's aspirations might be better evident within the economy. See Keith Hart, Jean-Louis Laville, and Antonio David Cattani (2010).

12. Climate remained Tetet's favorite issue, partly because the climate community was so warm and attuned to justice issues and partly because the SDGs just didn't have the charisma to excite people the way climate did.

13. Conversations on the possible incorporation of human rights within climate change negotiations were initiated in 2008 by the inclusion of climate change in a resolution by the United Nations Human Rights Council (Limon 2009). Since that time, this issue has acquired a lot of traction (see Rajamani 2010; Walbott and Schapper 2017).

14. See Harriet Thew, Lucie Middlemiss, and Jouni Paavola (2021) for a discussion on whether organizing youth participation within the UNFCCC has led to more youth access and authority within the process and how. They feel that although youth have more access into the space of the sessions, they do not necessarily have more access to organized initiatives and therefore still lack a meaningful impact on negotiations.

5. Accounting for Change in the Paris Agreement

1. The phrase "theory of change" was coined by Peter Drucker, who was a sociologist and management science theorist. He used the phrase in his 1954 book *The Practice of Management*. I am not sure that those who used this phrase associated it with Drucker and his theories of largely corporate management.

2. Whereas polyphony implies distinct, even contradictory, voices within the same domain but that can co-exist, heteroglossia suggests a situation in which these voices confront and conflict with one another (Bakhtin 2013).

3. See Daniel Klein et al. (2017) for a close legal reading and discussion of each element of the Paris Agreement as opposed to my literary and textual reading of it.

4. The fall from grace of science was indicated by the fact that the IPCC's most recent report on pathways to limiting temperature rise to only 1.5°C, commissioned by the COP, was sidelined by a bloc constituting the United States, Russia, and Saudi Arabia at COP24 in Poland in 2018. They ensured that the language in the final decision text for that COP should read that the report had been "taken note of" rather than "adopted" by the decision text coming out of that COP. At the same time, many have told me that while science may have served as the spur for the Rio meeting and that the IPCCC exists to support the Convention, giving the issue of climate change the support of the global scientific community, in practice science and scientists have not played a large part in the actual negotiations. As we saw in earlier chapters, activists emphasized fidelity to climate justice over upholding the authority of science, and Parties and blocs were largely led by national or group interests (Bradshaw, Van de Graaf, and Connolly 2019).

5. It is noteworthy that the larger UN already has a set of metrics for its sustainable development goals (SDGs), which were launched in 2015, in which work is underway to bring climate indices into closer conversation with social and economic goals.

6. A Thrice-Told Tale of Negotiations

1. I do not mark out any particular country, group of countries, or blocs as blockers within the process, as this figure mutated so much over the eight years that I have been studying this process. I have instead just pointed out when specific countries display the characteristics of being intransigent, uncooperative, behaving in a "rogue" fashion, etc.

2. Some useful texts on how to study negotiations in terms of their linguistic content and/or their process are those by Daniel Druckman (1997), Stefan Groth (2012), and Rudolf Vetschera, Sabine T. Koeszegi, and Michael Filzmoser (2021).

3. The ultimate objective of this Convention and any related legal instruments that the Conference of the Parties may adopt is to achieve, in accordance with the relevant provisions of the Convention, stabilization of greenhouse gas concentrations in the atmosphere at a level that would prevent dangerous anthropogenic interference with the climate system. Such a level should be achieved within a timeframe sufficient to allow ecosystems to adapt naturally to climate change, to ensure that food production is not threatened, and to enable economic development to proceed in a sustainable manner.

4. Other tasks relating to mitigation, such as determining a base year common to all, timeframes for reporting, public registries of NDCs, and response measures, were apportioned to the subsidiary bodies SBI or SBSTA.

5. See Johan Kaufman (1988), Ronald Walker (2004), and Joanna Depledge (2013) for a normative description of what is called conference diplomacy.

6. Drawing on the convention adopted within the process, media reportage, and policy and advocacy writings, I assimilate positions either verbally expressed or

submitted in writing by Party delegates to the countries or the blocs that they were representing. I apologize if it appears overly schematic, and thereby disconcerting, but this was how Party delegates sought to appear.

7. Also it is important to note, recalling Richard Kinley's words, that none of the negotiators, whether facilitators or Party delegates, were actually writing the texts themselves. The official text emerging in the negotiation rooms was the product of the personnel from the Secretariat, who were ever present, ever taking notes, and whose textual products came to carry the signature of the facilitators as their notes and later the contact group as their negotiated text. The Party views that were being submitted were, on the other hand, the products of Party delegates, and they showed considerable idiomatic variation. These submissions were largely in English, as their translation from the national languages of the Parties was not under the Secretariat's purview (Depledge 2013).

8. To remind the reader that the NDCs were an innovation of the Paris Agreement by which countries would report on what they planned to do to mitigate (ex ante), they would also provide reports after the fact on what they had done (ex poste), with these reports coming periodically (with start dates and timeframes still very much under discussion during the negotiations covered in this chapter). It was joined by two other innovations to ensure that the NDCs became more ambitious over time: One was the transparency framework (Article 13 of the Paris Agreement), and the second was the global stocktake (Article 14). The transparency framework was to apply to all reportage by Parties, and the global stocktake was to happen also every five years from the year of implementation of the Paris Rulebook to look comprehensively across all climate actions reported by Parties to see if they collectively brought the world to significant progress on checking temperature rise. The issue here was whether the discussion and final template for the NDCs should be similarly comprehensive as the transparency framework or the global stocktake or whether it should stay mitigation focused, as its mandate came from article of the Paris Agreement that was considered mitigation focused (see Van Asselt et al. 2016; Winkler, Mantlana, and Letete 2017; Bhushan and Rattani 2017; Ciplet et al. 2018; Northrop et al. 2018; Hermwille et al. 2019; Winkler 2020).

7. The House of Loss and Damage

1. One thing that is implicit and not often remarked upon is that loss and damage, as a notion, only has salience in a world in which there is already knowledge of climate change. It cannot be employed retroactively to describe environmental devastation before climate change was acknowledged by nation-states at the Rio Convention in 1992. So, however loss and damage is used, it is used in a time-restricted fashion and hinges on the question of whether there was prior knowledge of the possibility of loss and damage hanging over it, even if this question is not always asked or explored.

2. See https://unfccc.int/topics/adaptation-and-resilience/workstreams/loss-and -damage-ld/chronology-ld-workstream#eq-1.

3. See Roberts and Huq (2015) for an account of the emergence of WIM within the climate negotiations.

4. The distinction between extreme events and slow-onset events has also been hard fought, with developed countries within the process more open to considering helping with extreme events under the framework of humanitarian aid than to helping with slow-onset events, which committed them to long-term financing (Robinson et al. 2021).

5. Since writing this blog post, Saleem had been busy showing how solidarity payments could work. See https://www.iied.org/new-solidarity-funds-could -ringfence-finance-for-loss-damage.

6. See the 2017 annual Civil Society Equity Review.

7. See Martin L. Parry (2009) for efforts to assess the costs of adapting to climate change. The numbers are at the scale of the inordinate, making it hard to wrap one's mind around it.

8. The discussion on loss and damage related to climate change among scientists is equally lively, if not outright contentious (Schneider 2009). Experts in the IPCC have a very narrow understanding of the phrase, which is almost exclusively restricted to physical impacts, with limited reference to associated economic loss. This constrains how loss and damage is mediated to a wider public and acted upon (see Brysse et al. 2013; der Geest and Warner 2020).

Conclusion: The Gift of the Global South

1. Asad has since returned to organizing within the climate negotiations. He was front and center at the rallies and public events in Glasgow. DCJ created a huge coalition under the banner of climate justice of which many youth organizations and movements were a part.

Bibliography

Primary Documents

Parties. 2017. *Submissions at APA1-3.*

UNFCCC. 2016a. *Agenda and Annotations. Note by the Executive Secretary.*

———. 2016b. *APA 1 Agenda and Annotations. Note by the Executive Secretary.*

———. 2016c. *APA Scenario Note.*

———. 2016d. *Informal Note by the Co-Facilitators.*

———. 2016e. *Parties' Views Regarding Further Guidance in Relation to the Mitigation Section of Decision 1/CP.21.*

———. 2016f. *Report of the Conference of the Parties on Its Twenty-First Session, Held in Paris from 30 November to 13 December 2015.*

———. 2017a. *Agenda and Annotations. Note by the Executive Secretary.*

———. 2017b. *Agenda and Annotations. Note by the Executive Secretary.*

———. 2017c. *Informal Note by the Co-Facilitators.*

———. 2017d. *Preliminary Material in Preparation for the First Iteration of the Informal Note.*

———. 2017e. *Preliminary Material in Preparation for the First Iteration of the Informal Note.*

———. 2017f. *Reflections Note on the Third Part of the First Session of the Ad Hoc Working Group on the Paris Agreement.*

———. 2018a. *Agenda and Annotations. Note by the Executive Secretary.*

———. 2018b. *Agenda and Annotations. Note by the Executive Secretary.*

———. 2018c. *Agenda and Annotations. Revised Note by the Executive Secretary.*

———. 2018d. *Draft Text on APA 1.7 Agenda Item 3.*

———. 2018e. *Draft Text on APA1.7 Agenda Item 3.*

———. 2018f. *Joint Reflections Note by the Presiding Officers of the Ad Hoc Working Group on the Paris Agreement, the Subsidiary Body for Scientific and Technological Advice and the Subsidiary Body for Implementation.*

———. 2018g. *Navigation Tool by the Co-Facilitators.*

———. 2018h. *Reflections Note on the Fourth Part of the First Session of the Ad Hoc Working Group on the Paris Agreement.*

———. 2018i. *Tools Additional to and Based on the Information Notes Contained in the Annex to the Conclusions of the Ad Hoc Working Group on the Paris Agreement from the Fifth Part of Its First Session: Informal Document by the Co-Chairs (APA).*

———. 2019. *Decision 4/CMA.1. Further Guidance in Relation to the Mitigation Section of Decision 1/CP.21.*

Kyoto Protocol to the United Nations Framework Convention on Climate Change (1998).

Paris Agreement (2015).

United Nations Framework Convention on Climate Change (1992).

World Commission on Environment and Development, and G. H. Brundtland. 1987. *Our Common Future.* Oxford University Press.

Secondary Material

Abeysinghe, Achala Chandani. 2011a. "Helping UN Negotiators Protect the Poorest." *Reflect & React*, IIED, July 2011, https://www.iied.org/sites/default/files/pdfs/migrate/G03125.pdf.

Adelman, Sam. 2016. "Climate Justice, Loss and Damage, and Compensation for Small Island Developing States." *Journal of Human Rights and the Environment* 7(1): 32–53.

Adger, W. Neil, Suraje Dessai, Marisa Goulden, et al. 2009. "Are There Social Limits to Adaptation to Climate Change?" *Climatic Change* 93(3): 335–54.

Adger, W. Neil, Irene Lorenzoni, and Karen L. O'Brien. 2009. *Adapting to Climate Change: Thresholds, Values, Governance.* Cambridge University Press.

Adger, W. Neil, Jouni Paavola, Saleemul Huq, and Mary Jane Mace. 2006. *Fairness in Adaptation to Climate Change.* MIT Press.

Ajit, T. 2016. "Concerns over Mode of Work in APA Informal Consultation." *Third World Network.* https://twn.my/title2/climate/news/marrakech01/TWN_update9.pdf.

———. 2017. "Call for Strengthening Inter-Linkages in the Paris Agreement Work." *Third World Network.* https://www.twn.my/title2/climate/news/Bonn19/TWN_update10.pdf.

Ajit, T., and Meena Raman. 2016. "First Session of CMA Expected to Address Issue of "Homeless Matters" under Paris Agreement." *Third World Network.* https://twn.my/title2/climate/news/marrakech01/TWN_update17.pdf.

Alam, Shawkat, Sumudu Atapattu, Carmen G. Gonzalez, and Jona Razzaque. 2015. *International Environmental Law and the Global South*. Cambridge University Press.

Allan, Jen Iris. 2019. "Dangerous Incrementalism of the Paris Agreement." *Global Environmental Politics* 19(1): 4–11.

Amdur, Robert. 1977. "Global Distributive Justice: A Review Essay." *Journal of International Affairs* 31(1): 81–88.

Amin, Samir. 1977. "Self-Reliance and the New International Economic Order." *Monthly Review* 29(3): 1–21.

———. 1978. "Unequal Development: An Essay on the Social Formations of Peripheral Capitalism." *Science and Society* 42(2): 219–22.

Andresen, Steinar, and Shardul Agrawala. 2002. "Leaders, Pushers, and Laggards in the Making of the Climate Regime." *Global Environmental Change* 12(1): 41–51.

Anghie, Antony. 2001. "Colonialism and the Birth of International Institutions: Sovereignty, Economy, and the Mandate System of the League of Nations." *NYU Journal of International Law and Politics* 34:513–633.

———. 2006. "The Evolution of International Law: Colonial and Postcolonial Realities." *Third World Quarterly* 27(5): 739–53.

———. 2009. "Rethinking Sovereignty in International Law." *Annual Review of Law and Social Science* 5:291–310.

Ari, Izzet, and Ramazan Sari. 2017. "Differentiation of Developed and Developing Countries for the Paris Agreement." *Energy Strategy Reviews* 18:175–82.

Aronoff, Kate. 2021. *Overheated: How Capitalism Broke the Planet—and How We Fight Back*. Bold Type Books.

Artis, Amélie. 2017. "Social and Solidarity Finance: A Conceptual Approach." *Research in International Business and Finance* 39:737–49.

Asdal, Kristin. 2015. "What Is the Issue? The Transformative Capacity of Documents." *Distinktion: Scandinavian Journal of Social Theory* 16(1): 74–90.

Averchenkova, Alina, and Dimitri Zenghelis. 2018. "Pre-2020 Ambition on Climate Change: History, Status, Outlook." ecbi. https://ecbi.org/sites/default/files/Pre -2020%20Ambition.pdf.

Aykut, Stefan C. 2016. "Taking a Wider View on Climate Governance: Moving Beyond the 'Iceberg,' the 'Elephant,' and the 'Forest.'" *Wiley Interdisciplinary Reviews: Climate Change* 7(3): 318–28.

Baer, Hans A. 2015. "Towards an Anthropology of the Future: Visions of a Future World in the Era of Climate Change." In *Environmental Change and the World's Futures*, 17–32. Routledge.

Bailey, Sydney D. 1985. "Non-Official Mediation in Disputes: Reflections on Quaker Experience." *International Affairs (Royal Institute of International Affairs)* 61(2): 205–22.

Baillat, Alice. 2018. "From Vulnerability to Weak Power: Bangladesh in the Fight against Climate Change." *Revue Internationale et Strategique* 1:171–80.

Bakhtin, Mikhail. 2013. *Problems of Dostoevsky's Poetics*. University of Minnesota Press.

Barber, Karin. 2007. *The Anthropology of Texts, Persons, and Publics*. Cambridge University Press.

Barnett, Jon. 2008. "The Worst of Friends: OPEC and G-77 in the Climate Regime." *Global Environmental Politics* 8(4): 1–8.

Barrios, Roberto E. 2016. "Resilience: A Commentary from the Vantage Point of Anthropology." *Annals of Anthropological Practice* 40(1): 28–38.

Bateson, Gregory. 1958. *Naven: A Survey of the Problems Suggested by a Composite Picture of the Culture of a New Guinea Tribe Drawn from Three Points of View*. Stanford University Press.

Becker, Peter, and William Clark. 2001. *Little Tools of Knowledge: Historical Essays on Academic and Bureaucratic Practices*. University of Michigan Press.

Bendell, Jem, and Rupert Read. 2021a. *Deep Adaptation: Navigating the Realities of Climate Chaos*. John Wiley & Sons.

Bernardo, Cecília Silva, Gebru Jember Endalew, Thinley Namgyel, and Binyam Yakob Gebreyes. 2020a. "Least Developed Countries (LDCs)." In *Negotiating Climate Change Adaptation*, 61–71. Springer.

Bernstein, Steven F. 2001. *The Compromise of Liberal Environmentalism*. Columbia University Press.

Betzold, Carola. 2010 "Borrowing Power to Influence International Negotiations: AOSIS in the Climate Change Regime, 1990–1997." *Politics* 30(3): 131–48.

Bhushan, Chandra, and Vijeta Rattani. 2017. "Global Stocktake under the Paris Agreement." Policy brief, Center for Science and Environment.

Bidwai, Praful. 2012. "Climate Change, India, and the Global Negotiations." *Social Change* 42(3): 375–90.

Bierschenk, Thomas, and Jean-Pierre Olivier de Sardan. 2021. "The Anthropology of Bureaucracy and Public Administration." In *Oxford Research Encyclopedia of Politics*.

Blaxekjær, Lau Øfjord, Bård Lahn, Tobias Dan Nielsen, Lucia Green-Weiskel, and Fang Fang. 2020. "The Narrative Position of the Like-Minded Developing Countries in Global Climate Negotiations." In *Coalitions in the Climate Change Negotiations*, ed. C. Klöck, Paula Castro, Florian Weiler, and Lau Øfjord Blaxekjær, 113–35. Routledge, 2020.

Blaxekjær, Lau Øfjord, and Tobias Dan Nielsen. 2015. "Mapping the Narrative Positions of New Political Groups under the UNFCCC." *Climate Policy* 15(6): 751–66.

Bodansky, Daniel. 2001. "The History of the Global Climate Change Regime." *International Relations and Global Climate Change* 23(23): 505.

———. 2010. *The Art and Craft of International Environmental Law*. Harvard University Press.

———. 2016. "The Legal Character of the Paris Agreement." *Review of European, Comparative & International Environmental Law* 25(2): 142–50.

Boisard, Marcel A., Evgeny Chossudovsky, and Jacques Lemoine. 1998. *Multilateral Diplomacy: The United Nations System at Geneva; a Working Guide*. Kluwer Law International.

Bomzan, Prerna. 2018. "Developing Countries Call for Greater Balance in Negotiations." *Third World Network*. https://www.twn.my/title2/climate/bonn .news.21.htm.

Bomzan, Prerna, and Meena Raman. 2017. "APA Takes Stock of Progress towards Negotiating Text for Paris Agreement Implementation" *Third World Network*. https://www.twn.my/title2/climate/news/Bonn20/TWN_update8.pdf.

Borie, Maud, and Mike Hulme. 2015. "Framing Global Biodiversity: IPBES between Mother Earth and Ecosystem Services." *Environmental Science & Policy* 54:487–96.

Bose, Indrajit, and Meena Raman. 2016. "Battle of Interpretation over Paris Agreement Begins." *Third World Network*. https://twn.my/title2/climate/news /Bonn18/TWN_update1.pdf.

———. 2018. "Important Finance Decisions Adopted at Climate Talks." *Third World Network*. https://twn.my/title2/climate/news/katowice01/TWN_update15.pdf.

Bradshaw, Michael, Thijs Van de Graaf, and Richard Connolly. 2019. "Preparing for the New Oil Order? Saudi Arabia and Russia." *Energy Strategy Reviews* 26:100374.

Brenner, Reuven, and Gabrielle A. Brenner. 1990. *Gambling and Speculation: A Theory, a History, and a Future of Some Human Decisions*. Cambridge University Press.

Broberg, Morten, and Beatriz Martinez Romera. 2020. "Loss and Damage after Paris: More Bark Than Bite?" *Climate Policy* 20(6): 661–68.

Brysse, Keynyn, Naomi Oreskes, Jessica O'Reilly, and Michael Oppenheimer. 2013. "Climate Change Prediction: Erring on the Side of Least Drama?" *Global Environmental Change* 23(1): 327–37.

Burger, Michael, Jessica Wentz, and Radley Horton. 2020. "The Law and Science of Climate Change Attribution." *Columbia Journal of Environmental Law* 45:57.

Burkett, Maxine. 2014. "Loss and Damage." *Climate Law* 4(1–2): 119–30.

Calliari, Elisa. 2018. "Loss and Damage: A Critical Discourse Analysis of Parties' Positions in Climate Change Negotiations." *Journal of Risk Research* 21(6): 725–47.

Calliari, Elisa, Olivia Serdeczny, and Lisa Vanhala. 2020. "Making Sense of the Politics in the Climate Change Loss & Damage Debate." *Global Environmental Change* 64:102133.

Callison, Candis. 2014. *How Climate Change Comes to Matter: The Communal Life of Facts*. Duke University Press.

Carbon Brief. "Analysis: Which Countries Have Sent the Most Delegates to COP26?" Carbon Brief: Clear on Carbon. https://www.carbonbrief.org/analysis -which-countries-have-sent-the-most-delegates-to-cop26/.

Carlarne, Cinnamon P. 2010. "Reassessing the Role of Religion in Western Climate Change Decision-Making." In *Muslim and Christian Understanding*, 159–72: Springer.

Cavell, Stanley. 1994. *A Pitch of Philosophy*. Harvard University Press.

Chadwick, Michael J. 1994. *Negotiating Climate Change: The Inside Story of the Rio Convention*. Cambridge University Press.

Chan, Nicholas. 2021a. "'Special Circumstances' and the Politics of Climate Vulnerability: African Agency in the UN Climate Change Negotiations." *Africa Spectrum* 56(3): 314–32.

———. 2021b. "Beyond Delegation Size: Developing Country Negotiating Capacity and NGO 'Support' in International Climate Negotiations." *International Environmental Agreements: Politics, Law, and Economics* 21(2): 201–17.

Chasek, Pamela S. 2001. *Earth Negotiations: Analyzing Thirty Years of Environmental Diplomacy*. United Nations University Press.

Chasek, Pamela S., and David L. Downie. 2020. *Global Environmental Politics*. Routledge.

Chayes, Abram, and Antonia Handler Chayes. 1998. *The New Sovereignty: Compliance with International Regulatory Agreements*. Harvard University Press.

Christiansen, Lars, Anne Olhoff, and Thomas Dale. 2020. "Understanding Adaptation in the Global Stocktake." Independent Global Stocktake. https://www .climateworks.org/wp-content/uploads/2020/05/Understanding-Adaptation-in-the -Global-Stocktake_iGST_UNEP-DTU.pdf.

Cica, Natasha. 2016. "Lawyer Mathew Stilwell Tells Why He Is Always Looking Long Term." *Mercury Hobart Magazine*, 6 July. https://www.pressreader.com /australia/mercury-hobart-magazine/20160709/281569470051760.

Ciplet, David, Kevin M. Adams, Romain Weikmans, and J. Timmons Roberts. 2018. "The Transformative Capability of Transparency in Global Environmental Governance." *Global Environmental Politics* 18(3): 130–50.

Ciplet, David, and J. Timmons Roberts. 2017. "Climate Change and the Transition to Neoliberal Environmental Governance." *Global Environmental Change* 46:148–56.

Coates, Ta-Nehisi. 2015. "The Case for Reparations." In *The Best American Magazine Writing 2015*, 1–50. Columbia University Press.

Cons, Jason. 2019. "Delta Temporalities: Choked and Tangled Futures in the Sundarbans." *Ethnos*: 1–22.

Cook, Lindsey Fielder, and Isobel Edwards. 2017. "A Negotiator's Toolkit." Quaker United Nations Office. https://quno.org/sites/default/files/resources/QUNO%20 Negotiators%20Tookit_for%20web.pdf.

Cooper, Richard N. 2001. "The Kyoto Protocol: A Flawed Concept." *Environmental Law Reporter—News and Analysis* 31:11484.

Corporate Accountability. "COP25 Bankrolled by Big Polluters." https://www .corporateaccountability.org/resources/cop25sponsors/.

Crate, Susan A., and Mark Nuttall. 2009. *Anthropology and Climate Change: From Actions to Transformations*. Left Coast Press.

———. 2016. *Anthropology and Climate Change: From Actions to Transformations*. Routledge.

CSO Equity Review. 2018. "Equity and the Ambition Ratchet: Towards a Meaningful 2018 Facilitative Dialogue." doi:10.6084/m9.figshare5917408.

Dagnet, Yamide, Nathan Cogswell, Lorena Gonzalez, et al. 2021. "Challenging Climate Negotiations Deliver Limited Progress toward COP26." World Resource Institute. https://www.wri.org/insights/challenging-climate-negotiations-deliver -limited-progress-toward-cop26.

Danowski, Déborah, and Eduardo Viveiros de Castro. 2016. *The Ends of the World*. John Wiley & Sons.

Das, Veena. 2006. *Life and Words*. University of California Press.

Dasgupta, Susmita, M. D. Moqbul Hossain, Mainul Huq, and David Wheeler. 2016. "Facing the Hungry Tide: Climate Change, Livelihood Threats, and Household Responses in Coastal Bangladesh." *Climate Change Economics* 7(3): 1650007.

d'Aspremont, Jean. 2012. "Wording in International Law." *Leiden Journal of International Law* 25(3): 575–602.

de Águeda Corneloup, Inés, and Arthur P. J. Mol. 2014. "Small Island Developing States and International Climate Change Negotiations: The Power of Moral 'Leadership.'" *International Environmental Agreements: Politics, Law, and Economics* 14(3): 281–97.

De Longueville, Florence, Pierre Ozer, François Gemenne, et al. 2020. "Comparing Climate Change Perceptions and Meteorological Data in Rural West Africa to Improve the Understanding of Household Decisions to Migrate." *Climatic Change* 160(1): 123–41.

Depledge, Joanna. 2005. "Against the Grain: The United States and the Global Climate Change Regime." *Global Change, Peace & Security* 17(1): 11–27.

———. 2006. "The Opposite of Learning: Ossification in the Climate Change Regime." *Global Environmental Politics* 6(1): 1–22.

———. 2007. "A Special Relationship: Chairpersons and the Secretariat in the Climate Change Negotiations." *Global Environmental Politics* 7(1): 45–68.

———. 2008a. "Striving for No: Saudi Arabia in the Climate Change Regime." *Global Environmental Politics* 8(4): 9–35.

———. 2013a. *The Organization of Global Negotiations: Constructing the Climate Change Regime*. Routledge.

Depledge, Joanna, Miguel Saldivia, and Cristina Peñasco. 2022. "Glass Half Full or Glass Half Empty? The 2021 Glasgow Climate Conference." *Climate Policy* 22(2): 147–57.

Dessai, Suraje, and Emma Lisa Schipper. 2003. "The Marrakech Accords to the Kyoto Protocol: Analysis and Future Prospects." *Global Environmental Change* 13(2): 149–53.

Dimitrov, Radoslav S. 2016a. "The Paris Agreement on Climate Change: Behind Closed Doors." *Global Environmental Politics* 16(3): 1–11.

Doelle, Meinhard. 2014. "The Birth of the Warsaw Loss & Damage Mechanism: Planting a Seed to Grow Ambition?" *Carbon & Climate Law Review*: 35–45.

Doherty, Brian, and Timothy Doyle. 2013. *Environmentalism, Resistance, and Solidarity: The Politics of Friends of the Earth International.* Springer.

Dooley, Kate, Christian Holz, Sivan Kartha, et al. 2021. "Ethical Choices behind Quantifications of Fair Contributions under the Paris Agreement." *Nature Climate Change* 11(4): 300–5.

Doyle, Julie. 2009. "Climate Action and Environmental Activism: The Role of Environmental NGOs and Grassroots Movements in the Global Politics of Climate Change." In *Climate Change and the Media*, 106–13. Peter Lang.

Druckman, Daniel. 1997. "Negotiating in the International Context." In *Peacemaking in International Conflict: Methods and Techniques*, 81–124. Washington, DC: United States Institute of Peace Press.

Dubash, Navroz K. 2013. "The Politics of Climate Change in India: Narratives of Equity and Cobenefits." *Wiley Interdisciplinary Reviews: Climate Change* 4(3): 191–201.

Durham, Deborah Lynn, Jacqueline S. Solway, and American Anthropological Association. 2017. *Elusive Adulthoods: The Anthropology of New Maturities.* Indiana University Press.

DW. 2019. "Peruvian Farmer Takes on Energy Giant RWE." 6 December. https://www.dw.com/en/peruvian-farmer-takes-on-german-energy-giant-rwe/a-51546216.

ecbi. 2017. 2017 *Pre-Cop Training Workshop.* https://ecbi.org/publications/2017-pre-cop-training-workshop.

Eckersley, Robyn. 2012. "Moving Forward in the Climate Negotiations: Multilateralism or Minilateralism?" *Global Environmental Politics* 12(2): 24–42.

———. 2020. "Rethinking Leadership: Understanding the Roles of the US and China in the Negotiation of the Paris Agreement." *European Journal of International Relations* 26(4): 1178–202.

Edwards, Guy, and J. Timmons Roberts. 2015. *A Fragmented Continent: Latin America and the Global Politics of Climate Change.* MIT Press.

Edwards, Paul N. 2010. *A Vast Machine: Computer Models, Climate Data, and the Politics of Global Warming.* MIT Press.

Elliott, David, and Lindsey Fielder Cook. 2016. "Climate Justice and the Use of Human Rights Law in Reducing Greenhouse Gas Emissions." Quaker United Nations Office. https://quno.org/sites/default/files/resources/Climate%20Justice_August_2016.pdf.

Ellis, Jane, and Sara Moarif. 2015. "Identifying and Addressing Gaps in the UNFCCC Reporting Framework." OECD Climate Change Expert Group, Paper 7. https://www.oecd-ilibrary.org/docserver/5jm56w6f918n-en.pdf.

Engler, Mark, and Paul Engler. 2016. *This Is an Uprising: How Nonviolent Revolt Is Shaping the Twenty-First Century.* Bold Type Books.

Eslava, Luis, Michael Fakhri, and Vasuki Nesiah. 2017. *Bandung, Global History, and International Law: Critical Pasts and Pending Futures*. Cambridge University Press.

Eslava, Luis, and Sundhya Pahuja. 2020. "The State and International Law: A Reading from the Global South." *Humanity: An International Journal of Human Rights, Humanitarianism, and Development* 11(1): 118–38.

Falkner, Robert. 2016a. "The Paris Agreement and the New Logic of International Climate Politics." *International Affairs* 92(5): 1107–25.

Falzon, Danielle. 2021. "Ideal Delegation: How Institutional Privilege Silences 'Developing' Nations in the UN Climate Negotiations." *Social Problems*, August.

Favret-Saada, Jeanne. 2015. *The Anti-Witch*. Hau.

Forrester, Katrina, and Duncan Bell. 2019. "Reparations, History, and the Origins of Global Justice." *Empire, Race, and Global Justice*, ed. Duncan Bell, 22–51. Cambridge University Press.

Forsythe, David P., Roger A. Coate, and Kelly-Kate Pease. 2013. *The United Nations and Changing World Politics*. Westview.

Foucault, Michel. 2001. *Fearless Speech*. Semiotext(e)/MIT Press.

Freire, Paulo. 2013. "Pedagogy of the Oppressed." In *Curriculum Studies Reader E2*, 131–39. Routledge.

Friman, Mathias, and Björn-ola Linnér. 2008. "Technology Obscuring Equity: Historical Responsibility in UNFCCC Negotiations." *Climate Policy* 8(4): 339–54.

Fry, Ian. n.d. *"The Negotiating Process" for Negotiations Training Workshop to Strengthen the Capacity of Pacific Island Countries to Negotiate and Implement the International Biodiversity, Biosafety and Climate Change Instruments.* Downloaded March 18, 2019.

Gach, Evan. 2019. "Normative Shifts in the Global Conception of Climate Change: The Growth of Climate Justice." *Social Sciences* 8(1): 24.

Geden, Oliver. 2016. "The Paris Agreement and the Inherent Inconsistency of Climate Policymaking." *Wiley Interdisciplinary Reviews: Climate Change* 7(6): 790–97.

Genovese, Federica. 2020. *Weak States at Global Climate Negotiations*. Cambridge University Press.

Getachew, Adom. 2019. *Worldmaking after Empire*. Princeton University Press.

Gewirtzman, Jonathan, Sujay Natson, Julie-Anne Richards, et al. 2018. "Financing Loss and Damage: Reviewing Options under the Warsaw International Mechanism." *Climate Policy* 18(8): 1076–86.

Ghosh, Amitav. 2018. *The Great Derangement: Climate Change and the Unthinkable*. Penguin.

Girod, Bastien, and Thomas Flüeler. 2009. "Future IPCC Scenarios—Lessons Learned and Challenges to Scenario Building in Climate Change Policy." https://www.researchgate.net/publication/228370307_Future_IPCC_scenarios -lessons_learned_and_challenges_to_scenario_building_in_climate_change _policy.

Glaab, Katharina, Doris Fuchs, and Johannes Friederich. 2018. "Religious NGOs at the UNFCCC: A Specific Contribution to Global Climate Politics?" In *Religious NGOs at the United Nations*, 47–63. Routledge.

Goeminne, Gert, and Erik Paredis. 2010. "The Concept of Ecological Debt: Some Steps Towards an Enriched Sustainability Paradigm." *Environment, Development, and Sustainability* 12(5): 691–712.

Goodell, Jeff. 2015. "Obama Takes on Climate Change: The *Rolling Stone* Interview." *Rolling Stone*, 23 September.

Goso, Shiori. 2020. "Bangladesh LNG Power Plant Backed by \$645m in Japanese Lending." *Nikkei Asia*, 15 July, https://asia.nikkei.com/Business/Energy /Bangladesh-LNG-power-plant-backed-by-645m-in-Japanese-lending.

Groth, Stefan. 2012. *Negotiating Tradition—The Pragmatics of International Deliberations on Cultural Property*. Universitätsverlag Göttingen.

Gündoğdu, Ayten. 2014. "A Revolution in Rights: Reflections on the Democratic Invention of the Rights of Man." *Law, Culture, and the Humanities* 10(3): 367–79.

Gunningham, Neil. 2019. "Averting Climate Catastrophe: Environmental Activism, Extinction Rebellion, and Coalitions of Influence." *King's Law Journal* 30(2): 194–202.

Gupta, Joyeeta. 2014. *The History of Global Climate Governance*. Cambridge University Press.

Guyer, Jane I. 2004. *Marginal Gains: Monetary Transactions in Atlantic Africa*. Lewis Henry Morgan Lectures. University of Chicago Press.

Haddad, Lawrence James, John Hoddinott, Harold Alderman, and International Food Policy Research Institute. 1997. *Intrahousehold Resource Allocation in Developing Countries: Models, Methods, and Policy*. Johns Hopkins University Press.

Hadden, Jennifer. 2015. *Networks in Contention*. Cambridge University Press.

Haenn, Nora, Richard R. Wilk, and Allison Harnish. 2016. *The Environment in Anthropology: A Reader in Ecology, Culture, and Sustainable Living*. 2nd ed. New York University Press.

Hage, Ghassan. 2017. *Is Racism an Environmental Threat?* John Wiley & Sons.

Halkyer, Rene Orellana. 2021. "8 COP 21—Complaints and Negotiation: The Role of the Like-Minded Developing Countries Group (LMDC) and the Paris Agreement." In *Negotiating the Paris Agreement: The Insider Stories*, 160.

Halme-Tuomisaari, Miia. 2016. "Toward a Lasting Anthropology of International Law/Governance." *European Journal of International Law* 27(1): 235–43.

Hamilton, Clive. 2013. *Earthmasters: The Dawn of the Age of Climate Engineering*. Yale University Press.

Haque, Masroora, and Saleemul Huq. 2015. "Bangladesh and the Global Climate Debate." International Centre for Climate Change and Development. http:// www.icccad.net/wp-content/uploads/2015/12/Current-History-Bangladesh-and-the -Global-Climate-Debate-Haque-Huq.pdf.

Harper, Richard. 2009. *Inside the IMF*. Routledge.

Hart, Keith, Jean-Louis Laville, and Antonio David Cattani. 2010. *The Human Economy*. Polity.

Haug, Constanze, and Frans Berkhout. 2010. "Learning the Hard Way? European Climate Policy after Copenhagen." *Environment: Science and Policy for Sustainable Development* 52(3): 20–27.

Hénaff, Marcel. 2019. *The Philosophers' Gift*. Fordham University Press.

Hermwille, Lukas, Anne Siemons, Hannah Förster, and Louise Jeffery. 2019. "Catalyzing Mitigation Ambition under the Paris Agreement: Elements for an Effective Global Stocktake." *Climate Policy* 19(8): 988–1001.

Hernandez, Gleider I. 2012. "A Reluctant Guardian: The International Court of Justice and the Concept of 'International Community.'" *British Yearbook of International Law* 83(1): 13–60.

Heyward, Madeleine. 2007. "Equity and International Climate Change Negotiations: A Matter of Perspective." *Climate Policy* 7(6): 518–34.

Hickel, Jason. 2021. "The Anti-Colonial Politics of Degrowth." *Political Geography* 88:102404.

Hilton, Isabel, and Oliver Kerr. 2017. "The Paris Agreement: China's 'New Normal' Role in International Climate Negotiations." *Climate Policy* 17(1): 48–58.

Hirsch, Thomas. 2019. *Climate Finance for Addressing Loss and Damage: How to Mobilize Support for Developing Countries to Tackle Loss and Damage*. Brot für die Welt, Evangelisches Werk für Diakonie und Entwicklung eV.

Hjerpe, Mattias, and Björn-Ola Linnér. 2010. "Functions of COP Side-Events in Climate-Change Governance." *Climate Policy* 10(2): 167–80.

Hochstetler, Kathryn Ann. 2012. "The G-77, BASIC, and Global Climate Governance: A New Era in Multilateral Environmental Negotiations." *Revista Brasileira de Política Internacional* 55 (SPE): 53–69.

Hochstetler, Kathryn, and Manjana Milkoreit. 2014. "Emerging Powers in the Climate Negotiations: Shifting Identity Conceptions." *Political Research Quarterly* 67(1): 224–35.

Hoffmann, Matthew. 2016. "The Analytic Utility (and Practical Pitfalls) of Accountability." *Global Environmental Politics* 16(2): 22–32.

Holden, Emily, and Kalina Oroschakoff. 2017. "White House Coal Pitch Sparks Climate Outcry in Bonn." *Politico*, 13 November.

Honkonen, Tuula. 2009. *The Common but Differentiated Responsibility Principle in Multilateral Environmental Agreements: Regulatory and Policy Aspects*. Vol. 5. Kluwer Law International BV.

Hossain, Md Fahad, Danielle Falzon, M. Feisal Rahman, and Saleemul Huq. 2021. "Toward Climate Justice." In *Principles of Justice and Real-World Climate Politics*, ed. Sarah Kenehan and Corey Katz, 149–70. Rowman & Littlefield.

Huggel, Christian, Dáithí Stone, Maximilian Auffhammer, and Gerrit Hansen. 2013. "Loss and Damage Attribution." *Nature Climate Change* 3(8): 694–96.

Huggel, Christian, Dáithí Stone, Hajo Eicken, and Gerrit Hansen. 2015. "Potential and Limitations of the Attribution of Climate Change Impacts for Informing Loss and Damage Discussions and Policies." *Climatic Change* 133(3): 453–67.

Hughes, David McDermott. 2017. *Energy without Conscience: Oil, Climate Change, and Complicity.* Duke University Press.

Hulme, Mike. 2014. "Attributing Weather Extremes to 'Climate Change': A Review." *Progress in Physical Geography* 38(4): 499–511.

———. 2016. *Weathered: Cultures of Climate.* Sage.

Hulme, Mike, Saffron J. O'Neill, and Suraje Dessai. 2011. "Is Weather Event Attribution Necessary for Adaptation Funding?" *Science* 334 (6057): 764–65.

Huq, Saleemul. 2013. "A House with Many Rooms: Addressing Loss and Damage from Climate Change." IIED, 4 September. https://www.iied.org/house-many -rooms-addressing-loss-damage-climate-change.

Huq, Saleemul, Erin Roberts, and Adrian Fenton. 2013. "Loss and Damage." *Nature Climate Change* 3(11): 947–49.

Ikeme, Jekwu. 2003. "Equity, Environmental Justice, and Sustainability: Incomplete Approaches in Climate Change Politics." *Global Environmental Change* 13(3): 195–206.

Isenhour, Cindy, Jessica O'Reilly, and Heather Yocum. 2016. "Introduction to Special Theme Section: Accounting for Climate Change: Measurement, Management, Morality, and Myth." *Human Ecology* 44(6): 647–54.

Islam, Md Mofakkarul. 2022. "Distributive Justice in Global Climate Finance— Recipients' Climate Vulnerability and the Allocation of Climate Funds." *Global Environmental Change* 73:102475.

Jacob, Merle. 1994. "Toward a Methodological Critique of Sustainable Development." *Journal of Developing Areas* 28(2): 237–52.

Jaeger, Carlo C., Klaus Hasselmann, Gerd Leipold, et al. 2012. "Reframing the Problem of Climate Change." In *From Zero-Sum Game to Win-Win Solutions.* Earthscan.

James, Rachel, Friederike Otto, Hannah Parker, et al. 2014. "Characterizing Loss and Damage from Climate Change." *Nature Climate Change* 4(11): 938–39.

James, Rachel A., Richard G. Jones, Emily Boyd, et al. 2019. "Attribution: How Is It Relevant for Loss and Damage Policy and Practice?" In *Loss and Damage from Climate Change,* 113–54. Springer.

Jamieson, Dale. 2014. *Reason in a Dark Time: Why the Struggle against Climate Change Failed—and What It Means for Our Future.* Oxford University Press.

Jefford, Stuart, and Dan Hamza-Goodacre. 2013. "Supporting International Climate Negotiators: Lessons from CDKN." Climate and Development Knowledge Network, August. https://cdkn.org/sites/default/files/files/CDKN_Working-Paper _Negotiations_web-final2.pdf.

Jepsen, Henrik, Magnus Lundgren, Kai Monheim, and Hayley Walker. 2021. *Negotiating the Paris Agreement.* Cambridge University Press.

Jézéquel, Aglaé, Pascal Yiou, and Jean-Paul Vanderlinden. 2019. "Comparing Scientists' and Delegates' Perspectives on the Use of Extreme Event Attribution for Loss and Damage." *Weather and Climate Extremes* 26:100231.

Johns, Fleur. 2013. *Non-Legality in International Law: Unruly Law.* Cambridge University Press.

Kartiki, Katha. 2011. "Climate Change and Migration: A Case Study from Rural Bangladesh." *Null* 19(1): 23–38.

Kasa, Sjur, Anne T. Gullberg, and Gørild Heggelund. 2008. "The Group of 77 in the International Climate Negotiations: Recent Developments and Future Directions." *International Environmental Agreements: Politics, Law, and Economics* 8(2): 113–27.

Kaufman, Johan. 1988. *Conference Diplomacy: An Introductory Analysis.* M. Nijhoff.

Keck, Margaret E., and Kathryn Sikkink. 1998. *Activists beyond Borders: Advocacy Networks in International Politics.* Cornell University Press.

Kemp, Luke. 2016. "Framework for the Future? Exploring the Possibility of Majority Voting in the Climate Negotiations." *International Environmental Agreements: Politics, Law, and Economics* 16(5): 757–79.

———. 2018. "A Systems Critique of the 2015 Paris Agreement on Climate." In *Pathways to a Sustainable Economy,* 25–41. Springer.

Klein, Daniel, María Pía Carazo, Meinhard Doelle, et al. 2017. *The Paris Agreement on Climate Change: Analysis and Commentary.* Oxford University Press.

Klein, Naomi. 2015. *This Changes Everything: Capitalism vs. the Climate.* Simon and Schuster.

Klinsky, Sonja, Timmons Roberts, Saleemul Huq, et al. 2017. "Why Equity Is Fundamental in Climate Change Policy Research." *Global Environmental Change* 44:170–73.

Klöck, Carola, Paula Castro, Florian Weiler, and Lau Øfjord Blaxekjær, eds. 2020. *Coalitions in the Climate Change Negotiations.* Routledge.

Koh, Harold Hongju. 1996. "Why Do Nations Obey International Law?" *Yale Law Journal* 106:2599.

Kreienkamp, Julia, and Lisa Vanhala. 2017. "Climate Change Loss and Damage." *Global Governance Institute:* 1–28.

Kuyper, Jonathan W., and Karin Bäckstrand. 2016. "Accountability and Representation: Nonstate Actors in UN Climate Diplomacy." *Global Environmental Politics* 16(2): 61–81.

Lanchbery, John, and David Victor. 1995. "The Role of Science in the Global Climate Negotiations." *Green Globe Yearbook:* 29–39.

Lawrence, Mark G., and Stefan Schäfer. 2019. "Promises and Perils of the Paris Agreement." *Science* 364 (6443): 829–30.

Lawrence, Peter, and Daryl Wong. 2017. "Soft Law in the Paris Climate Agreement: Strength or Weakness?" *Review of European, Comparative, and International Environmental Law* 26 (3): 276–86.

Levy, David L., and Daniel Egan. 1998. "Capital Contests: National and Transnational Channels of Corporate Influence on the Climate Change Negotiations." *Politics and Society* 26 (3): 337–61.

Limon, Marc. 2009. "Human Rights and Climate Change: Constructing a Case for Political Action." *Harvard Environmental Law Review* 33:439.

Livingston, Jasmine E., and Markku Rummukainen. 2020. "Taking Science by Surprise: The Knowledge Politics of the IPCC Special Report on 1.5 Degrees." *Environmental Science and Policy* 112:10–16.

Lloyd, Elisabeth A., and Theodore G. Shepherd. 2021. "Climate Change Attribution and Legal Contexts: Evidence and the Role of Storylines." *Climatic Change* 167(3): 1–13.

Lusama, Tafue Molu. 2017. "A Letter from . . . Tuvalu." *Reform*, June. https://www .reform-magazine.co.uk/2017/05/a-letter-from-tuvalu/.

Lusk, Greg. 2017. "The Social Utility of Event Attribution: Liability, Adaptation, and Justice-Based Loss and Damage." *Climatic Change* 143(1): 201–12.

Lydon, Anthony F., M. A. Boisard, and E. M. Chossudovsky. 1998. "The Making of a United Nations Meeting." In *Multilateral Diplomacy: The United Nations System at Geneva: A Working Guide*, 149–60. Kluwer.

Lynas, Mark. 2008. *Six Degrees: Our Future on a Hotter Planet*. National Geographic Books.

Mace, Mary Jane, and Roda Verheyen. 2016. "Loss, Damage, and Responsibility after COP 21: All Options Open for the Paris Agreement." *Review of European, Comparative, and International Environmental Law* 25(2): 197–214.

MacNeil, Robert, and Matthew Paterson. 2020. "Trump, US Climate Politics, and the Evolving Pattern of Global Climate Governance." *Global Change, Peace, and Security* 32(1): 1–18.

Marjanac, Sophie, and Lindene Patton. 2018. "Extreme Weather Event Attribution Science and Climate Change Litigation: An Essential Step in the Causal Chain?" *Journal of Energy and Natural Resources Law* 36(3): 265–98.

Marquardt, Jens. 2017. "Conceptualizing Power in Multi-Level Climate Governance." *Journal of Cleaner Production* 154:167–75.

Marsden, William. 2011. *Fools Rule: Inside the Failed Politics of Climate Change*. Vintage Canada.

Martinez-Alier, Joan, Leah Temper, Daniela Del Bene, et al. 2016. "Is There a Global Environmental Justice Movement?" *Journal of Peasant Studies* 43(3): 731–55.

Mathur, Nayanika. 2017. "The Task of the Climate Translator." *Economic and Political Weekly* 52(31).

Matthews, H. Damon. 2016. "Quantifying Historical Carbon and Climate Debts among Nations." *Nature Climate Change* 6(1): 60–64.

Mauss, Marcel. 2002. *The Gift: The Form and Reason for Exchange in Archaic Societies*. Routledge.

Mayer, Benoit. 2016. "Human Rights in the Paris Agreement." *Climate Law* 6(1–2): 109–17.

McLaren, Duncan. 1992. "Friends of the Earth." *Planning Practice and Research* 7(3): 43–47.

McNamara, Karen Elizabeth. 2014. "Exploring Loss and Damage at the International Climate Change Talks." *International Journal of Disaster Risk Science* 5(3): 242–46.

Merry, Sally Engle. 2006. "Anthropology and International Law." *Annual Review of Anthropology* 35:99–116.

———. 2007. "International Law and Sociolegal Scholarship: Toward a Spatial Global Legal Pluralism." In A. Sarat, ed., *Special Issue: Law and Society Reconsidered (Studies in Law, Politics and Society)*: 149–68.

———. 2011. "Measuring the World: Indicators, Human Rights, and Global Governance." *Current Anthropology* 52(S3): S83–S95.

Mertz, Elizabeth. 2017. "Language, Law, and Social Meanings: Linguistic/Anthropological Contributions to the Study of Law." In *Legal Theory and the Social Sciences*, 361–93. Routledge.

Meyer, Robinson. 2015a. "Is Hope Possible after the Paris Agreement?" *Atlantic*, December 12.

Meyer, Robinson. 2015b. "A Reader's Guide to the Paris Agreement." *Atlantic*, December 16.

Michaelowa, Katharina, and Axel Michaelowa. 2012. "India as an Emerging Power in International Climate Negotiations." *Climate Policy* 12(5): 575–90.

Mignolo, Walter D. 2011. "The Global South and World Dis/Order." *Journal of Anthropological Research* 67(2): 165–88.

Milkoreit, Manjana. 2015. "Hot Deontology and Cold Consequentialism—an Empirical Exploration of Ethical Reasoning among Climate Change Negotiators." *Climatic Change* 130(3): 397–409.

———. 2017. *Mindmade Politics: The Cognitive Roots of International Climate Governance*. MIT Press.

Milkoreit, Manjana, and Kate Haapala. 2017a. *Designing the Global Stocktake: A Global Governance Innovation*. https://www.c2es.org/document/designing-the-global-stocktake-a-global-governance-innovation/.

Miller, Marian A. L. 1995. *The Third World in Global Environmental Politics*. Lynne Rienner.

Milne, Markus J., and Suzana Grubnic. 2011. "Climate Change Accounting Research: Keeping It Interesting and Different." *Accounting, Auditing, and Accountability* 24(8): 948–77.

Mitchell, Timothy. 2011. *Carbon Democracy: Political Power in the Age of Oil*. Verso.

Molder, Amanda L., Alexandra Lakind, Zoe E. Clemmons, et al. 2021. "Framing the Global Youth Climate Movement: A Qualitative Content Analysis of Greta Thunberg's Moral, Hopeful, and Motivational Framing on Instagram." *International Journal of Press/Politics* 27(3): 668–95.

Moore, Jason. 2015. *Capitalism in the Web of Life: Ecology and the Accumulation of Capital*. Verso.

Moyn, Samuel. 2018. *Not Enough: Human Rights in an Unequal World*. Belknap.

Moyn, Samuel, M. O'Brien, J. Isaac, et al. 2013. "The Political Origins of Global Justice." Lecture at the Center for the Humanities, Wesleyan University.

Müller, Birgit. 2013. *The Gloss of Harmony: The Politics of Policy-Making in Multilateral Organisations*. Pluto.

Murray, Brian C. 2000a. "Carbon Values, Reforestation, and 'Perverse' Incentives under the Kyoto Protocol: An Empirical Analysis." *Mitigation and Adaptation Strategies for Global Change* 5(3): 271–95.

Nader, Laura. 1969. "Up the Anthropologist: Perspectives Gained from Studying Up." In *Reinventing Anthropology*, ed. Dell Hymes, 284–311. Random House.

Nel, Adrian. 2015. "The Choreography of Sacrifice: Market Environmentalism, Biopolitics, and Environmental Damage." *Geoforum* 65:246–54.

Nelson, Alondra. 2016. *The Social Life of DNA: Race, Reparations, and Reconciliation after the Genome*. Beacon.

Niezen, Ronald, and Maria Sapignoli, eds. 2017. *Palaces of Hope: The Anthropology of Global Organizations*. Cambridge Studies in Law and Society. Cambridge University Press.

Nisbet, Matthew C. 2009. "Communicating Climate Change: Why Frames Matter for Public Engagement." *Environment: Science and Policy for Sustainable Development* 51(2): 12–23.

Nishat, Ainun, Nandan Mukherjee, Erin Roberts, et al. 2013. "A Range of Approaches to Address Loss and Damage from Climate Change Impacts in Bangladesh." BRAC University, Dhaka, Bangladesh.

Northrop, Eliza, Yamide Dagnet, Niklas Höhne, et al. 2018. "Achieving the Ambition of Paris: Designing the Global Stocktake." World Resources Institute (WRI), Washington, DC. https://www.wri.org/research/achieving-ambition-paris-designing-global-stocktake.

Obama, Barack. "Remarks by President Obama at the First Session of COP21." https://obamawhitehouse.archives.gov/the-press-office/2015/11/30/remarks-president-obama-first-session-cop21.

Obergassel, Wolfgang, Christof Arens, Christiane Beuermann, et al. 2019. "Time for Action—Blocked and Postponed. A First Assessment of COP25 in Madrid." https://www.researchgate.net/publication/338043558_Time_for_Action_-_Blocked_and_Postponed_A_first_assessment_of_COP25_in_Madrid.

———. 2020. "COP25 in Search of Lost Time for Action: An Assessment of the Madrid Climate Conference." *Carbon & Climate Law Review* 14(1): 3–17.

Ohdedar, Birsha. 2016. "Loss and Damage from the Impacts of Climate Change: A Framework for Implementation." *Nordic Journal of International Law* 85(1): 1–36.

Oliver-Smith, Anthony. 2016. "Contested Concepts in the Anthropology of Climate Change." In *Routledge Handbook of Environmental Anthropology*, 206–18. Routledge.

O'Reilly, Jessica, Cindy Isenhour, Pamela McElwee, et al. 2020a. "Climate Change: Expanding Anthropological Possibilities." *Annual Review of Anthropology* 49:13–29.

Ourbak, Timothée, and Alexandre K. Magnan. 2018. "The Paris Agreement and Climate Change Negotiations: Small Islands, Big Players." *Regional Environmental Change* 18(8): 2201–7.

Paprocki, Kasia. 2015a. "Anti-Politics of Climate Change." *Himal Southasian* 28(3): 54–64.

———. 2015b. "Anti-Politics of Climate Change." *Himal Southasian* 28(3): 54–64.

———. 2018. "Threatening Dystopias: Development and Adaptation Regimes in Bangladesh." *Annals of the American Association of Geographers* 108(4): 955–73.

———. 2021. *Threatening Dystopias: The Global Politics of Climate Change Adaptation in Bangladesh*. Cornell University Press.

———. 2022. "On Viability: Climate Change and the Science of Possible Futures." *Global Environmental Change* 73:102487. https://www.sciencedirect.com/science /article/pii/S0959378022000255.

Park, Susan, and Teresa Kramarz. 2019. *Global Environmental Governance and the Accountability Trap*. MIT Press.

Parker, Charles F., and Christer Karlsson. 2018. "The UN Climate Change Negotiations and the Role of the United States: Assessing American Leadership from Copenhagen to Paris." *Environmental Politics* 27(3): 519–40.

Parks, Bradley C., and J. Timmons Roberts. 2010. "Climate Change, Social Theory, and Justice." *Theory, Culture, and Society* 27(2–3): 134–66.

Parry, Martin L., Nigel Arnell, Pam Berry, David Dodman, Samuel Fankhauser, Chris Hope, Sari Kovats, Robert Nicholls, David Satterthwaite, Richard Tiffin, and Tim Wheeler. 2009. "Assessing the Costs of Adaptation to Climate Change: A Review of the UNFCCC and Other Recent Estimates." International Institute for Environment and Development. https://www.iied.org/sites/default/files/pdfs /migrate/11501IIED.pdf.

Patnaik, Utsa, and Prabhat Patnaik. 2016. *A Theory of Imperialism*. Columbia University Press.

Pearce, Fred. 1991. *Green Warriors the People and the Politics behind the Environmental Revolution*. Bodley Head.

———. 2007. *With Speed and Violence: Why Scientists Fear Tipping Points in Climate Change*. Beacon.

Piotrowski, Ryszard. 2011. "The Importance of Preamble in Constitutional Court Jurisprudence." *Acta Juridica Hungarica* 52(1): 29–39.

Poovey, Mary. 1998. *A History of the Modern Fact: Problems of Knowledge in the Sciences of Wealth and Society*. University of Chicago Press.

Pope Francis. 2015. *Laudato Si': On Care for Our Common Home*. http://www .vatican.va/content/francesco/en/encyclicals/documents/papa-francesco_20150524 _enciclica-laudato-si.html.

Pottier, Antonin, Aurélie Méjean, Olivier Godard, et al. 2017. "A Survey of Global Climate Justice: From Negotiation Stances to Moral Stakes and Back." *International Review of Environmental and Resource Economics* 11(1): 1–53.

Pouliot, Vincent. 2016. *International Pecking Orders: The Politics and Practice of Multilateral Diplomacy*. Cambridge University Press.

Prashad, Vijay. 2013. *The Poorer Nations: A Possible History of the Global South*. Verso.

Prigogine, Ilya, and Isabelle Stengers. 2018. *Order out of Chaos: Man's New Dialogue with Nature*. Verso.

Qi, Xinran. 2011. "The Rise of BASIC in UN Climate Change Negotiations." *South African Journal of International Affairs* 18(3): 295–318.

Radin, Margaret Jane. 1993. "Compensation and Commensurability." *Duke Law Journal* 43:56.

Rahman, Mofizur. 2018. "Climate Change Journalism in Bangladesh. Professional Norms and Attention in Newspaper Coverage of Climate Change." Thesis, University of Bergen, Norway. http://hdl.handle.net/1956/18924.

Rajagopal, Balakrishnan. 2003. *International Law from Below: Development, Social Movements, and Third World Resistance*. Cambridge University Press.

Rajamani, Lavanya. 2010a. "The Increasing Currency and Relevance of Rights-Based Perspectives in the International Negotiations on Climate Change." *Journal of Environmental Law* 22(3): 391–429.

———. 2014. "The Warsaw Climate Negotiations: Emerging Understandings and Battle Lines on the Road to the 2015 Climate Agreement." *International and Comparative Law Quarterly* 63(3): 721–40.

———. 2016. "Ambition and Differentiation in the 2015 Paris Agreement: Interpretative Possibilities and Underlying Politics." *International and Comparative Law Quarterly* 65(2): 493–514.

Rajão, Raoni and Tiago Duarte. 2018. "Performing Postcolonial Identities at the United Nations' Climate Negotiations." *Postcolonial Studies* 21(3): 364–78.

Raman, Meena. 2016. "The Climate Change Battle in Paris: An Initial Analysis of the Paris COP21 and the Paris Agreement." *Economic and Political Weekly*, January 9.

———. 2018a. "Decisions for Implementation of Paris Agreement Adopted." *Third World Network*. https://www.twn.my/title2/climate/news/katowice01/TWN _update13.pdf.

———. 2018b. "The Key Decisions on the Paris Agreement Implementation Rules." *Third World Network*. https://www.twn.my/title2/climate/news/katowice01/TWN _update14.pdf.

Raman, Meena, and Indrajit Bose. 2018. "Difficulties Expected at Year End Climate Talks." *Third World Network*. https://www.twn.my/title2/resurgence/2018/331-332 /eco1.htm.

Raman, Meena, and T. Ajit. 2017. "Issues Facing the Bonn Climate Talks." *Third World Network*. https://www.twn.my/title2/climate/bonn.news.19.htm.

Rietig, Katharina. 2014. "'Neutral' Experts? How Input of Scientific Expertise Matters in International Environmental Negotiations." *Policy Sciences* 47(2): 141–60.

Riles, Annelise. 1999a. "Global Designs: The Aesthetics of International Legal Practice." *Proceedings of the Annual Meeting (American Society of International Law)* 93:28–34.

———. 1999b. "Models and Documents: Artifacts of International Legal Knowledge." *International and Comparative Law Quarterly* 48(4): 805–25.

———. 2000. *The Network Inside Out.* University of Michigan Press.

———. 2001. "Encountering Amateurism: John Henry Wigmore and the Uses of American Formalism." Northwestern Public Law Research Paper no. 00-6.

Roberts, Erin, and Saleemul Huq. 2015. "Coming Full Circle: The History of Loss and Damage under the UNFCCC." *International Journal of Global Warming* 8(2): 141–57.

Roberts, J. Timmons. 2011. "Multipolarity and the New World (Dis)Order: US Hegemonic Decline and the Fragmentation of the Global Climate Regime." *Global Environmental Change* 21(3): 776–84.

Roberts, J. Timmons, and Romain Weikmans. 2017a. "Postface: Fragmentation, Failing Trust, and Enduring Tensions over What Counts as Climate Finance." *International Environmental Agreements: Politics, Law, and Economics* 17(1): 129–37.

Robinson, Meyer. 2015. "A Reader's Guide to the Paris Agreement." *Atlantic,* December 16.

Robinson, Stacy-ann, Mizan Khan, J. Timmons Roberts, et al. 2021. "Financing Loss and Damage from Slow Onset Events in Developing Countries." *Current Opinion in Environmental Sustainability* 50:138–48.

Rogelj, J., K. Jiang, J. Lowe, et al. 2015. "The Importance of Pre-2020 Action." https://pure.iiasa.ac.at/id/eprint/14925/.

Roger, Charles, and Satishkumar Belliethathan. 2016. "Africa in the Global Climate Change Negotiations." *International Environmental Agreements: Politics, Law, and Economics* 16(1): 91–108.

Rollosson, Natabara. 2010. "The United Nations Development Programme (UNDP) Working with Faith Representatives to Address Climate Change: The Two Wings of Ethos and Ethics." *CrossCurrents* 60(3): 419–31.

Schalatek, Liane, Neil Bird, and Jessica Brown. 2010a. "Where's the Money? The Status of Climate Finance Post-Copenhagen." https://www.boell.org/en/navigation/climate-energy-climate-finance-post-copenhagen-8706.html.

Schipper, E. Lisa F. 2006. "Conceptual History of Adaptation in the UNFCCC Process." *Review of European Community and International Environmental Law* 15(1): 82–92.

Schneider, Lambert Richard. 2011. "Perverse Incentives under the CDM: An Evaluation of HFC-23 Destruction Projects." *Climate Policy* 11(2): 851–64.

Schneider, Lambert, Maosheng Duan, Robert Stavins, et al. 2019. "Double Counting and the Paris Agreement Rulebook." *Science* 366(6462): 180–83.

Schneider, Stephen H. 2009. *Science as a Contact Sport: Inside the Battle to Save Earth's Climate*. National Geographic Books.

Schroeder, Heike, and Heather Lovell. 2012. "The Role of Non-Nation-State Actors and Side Events in the International Climate Negotiations." *Climate Policy* 12(1): 23–37.

Seaman, John A., Gary E. Sawdon, James Acidri, et al. 2014. "The Household Economy Approach. Managing the Impact of Climate Change on Poverty and Food Security in Developing Countries." *Climate Risk Management* 4:59–68.

Seddiky, Md Assraf, Helen Giggins, and Thayaparan Gajendran. 2020. "International Principles of Disaster Risk Reduction Informing NGOs Strategies for Community Based DRR Mainstreaming: The Bangladesh Context." *International Journal of Disaster Risk Reduction* 48:101580.

Selcer, Perrin. 2018. *The Postwar Origins of the Global Environment*. Columbia University Press.

Sen, Amartya. 1982. *Poverty and Famines: An Essay on Entitlement and Deprivation*. Oxford University Press.

Seo, S. Niggol. 2019. "The Green Climate Fund: History, Institution, Pledges, Investment Criteria." In *The Economics of Global Allocations of the Green Climate Fund*, 35–65. Springer.

Serdeczny, Olivia, Eleanor Waters, and Sander Chan. 2016. "Non-Economic Loss and Damage in the Context of Climate Change: Understanding the Challenges." Discussion paper. https://www.idos-research.de/uploads/media/DP_3.2016.pdf.

Siebert, Sabina, Fiona Wilson, and John RA Hamilton. 2017. "'Devils May Sit Here': The Role of Enchantment in Institutional Maintenance." *Academy of Management Journal* 60(4): 1607–32.

Siperstein, Stephen, Shane Hall, and Stephanie LeMenager. 2017. *Teaching Climate Change in the Humanities*. Routledge.

Soll, Jacob. 2014. *The Reckoning: Financial Accountability and the Rise and Fall of Nations*. Basic Books.

Spash, Clive L. 2016. "This Changes Nothing: The Paris Agreement to Ignore Reality." *Globalizations* 13(6): 928–33.

Stabinsky, Doreen, and Juan P. Hoffmaister. 2015. "Establishing Institutional Arrangements on Loss and Damage under the UNFCCC: The Warsaw International Mechanism for Loss and Damage." *International Journal of Global Warming* 8(2): 295–318.

Stadelmann, Martin, T. J. Roberts, and Saleemul Huq. 2010. "Baseline for Trust: Defining 'New and Additional' Climate Funding." *IIED Briefing Papers*: 1–4.

Streck, Charlotte, Moritz von Unger, and Sandra Greiner. 2020. "COP 25: Losing Sight of (Raising) Ambition." *Journal for European Environmental and Planning Law* 17(2): 136–60.

Surminski, Swenja, and Ana Lopez. 2015. "Concept of Loss and Damage of Climate Change—a New Challenge for Climate Decision-Making? A Climate Science Perspective." *Climate and Development* 7(3): 267–77.

Switzer, Jacqueline Vaughn. 2003. *Environmental Activism: A Reference Handbook.* ABC-CLIO.

Tarikul Islam, Mohammad. 2018. "Climate Negotiations: How Does Bangladesh Fare?" South Asia@ LSE. https://blogs.lse.ac.uk/southasia/2018/03/31/climate-negotiations-how-does-bangladesh-fare/.

Thew, Harriet, Lucie Middlemiss, and Jouni Paavola. 2020. "'Youth Is Not a Political Position': Exploring Justice Claims-Making in the UN Climate Change Negotiations." *Global Environmental Change* 61:102036.

———. 2021. "Does Youth Participation Increase the Democratic Legitimacy of UNFCCC-Orchestrated Global Climate Change Governance?" *Environmental Politics* 30(6): 873–94.

Thorsen, Dag Einar. 2010. "The Neoliberal Challenge. What Is Neoliberalism?" *Contemporary Readings in Law and Social Justice* 2(2): 188–214.

Tomz, Michael. 2012. *Reputation and International Cooperation*: Princeton University Press.

Toole, Stephanie, Natascha Klocker, and Lesley Head. 2016. "Re-Thinking Climate Change Adaptation and Capacities at the Household Scale." *Climatic Change* 135(2): 203–9.

Touval, Saadia. 1994. "Why the UN Fails." *Foreign Affairs* 73:44.

Toye, John. 2014. "Assessing the G77: 50 Years after UNCTAD and 40 Years after the NIEO." *Third World Quarterly* 35(10): 1759–74.

Umemiyaa, Chisa, Eri Ikedab, Veronika Gukovac, et al. 2021. "Understanding Support Needed for Climate Mitigation and Adaptation in Developing Countries from the National Reporting Under the UNFCCC." https://www.iges.or.jp/en/publication_documents/pub/workingpaper/en/11596/Support+needed_Umemiya+et+al_IGES+WP_final.pdf.

UN News. 2015. "Bangladeshi Prime Minister Wins UN Environment Prize for Leadership on Climate Change." September. https://news.un.org/en/story/2015/09/508702.

Urpelainen, Johannes, and Thijs Van de Graaf. 2018. "United States Non-Cooperation and the Paris Agreement." *Climate Policy* 18(7): 839–51.

Van Asselt, Harro, Romain Weikmans, J. Timmons Roberts, et al. 2016a. "Transparency of Action and Support under the Paris Agreement." SSRN 2859151.

van der Geest, Kees, and Koko Warner. 2015. "Loss and Damage from Climate Change: Emerging Perspectives." *International Journal of Global Warming* 8(2): 133–40.

———. 2020. "Loss and Damage in the IPCC Fifth Assessment Report (Working Group II): A Text-Mining Analysis." *Climate Policy* 20(6): 729–42.

Vanhala, Lisa, and Cecilie Hestbaek. 2016. "Framing Climate Change Loss and Damage in UNFCCC Negotiations." *Global Environmental Politics* 16(4): 111–29.

Verchick, Robert R. M. 2018. "Can 'Loss and Damage' Carry the Load?" *Philosophical Transactions of the Royal Society A: Mathematical, Physical, and Engineering Sciences* 376(2119): 20170070.

Verheyen, Roda. 2015. "Loss and Damage Due to Climate Change: Attribution and Causation—Where Climate Science and Law Meet." *International Journal of Global Warming* 8(2): 158–69.

Vetschera, Rudolf, Sabine T. Koeszegi, and Michael Filzmoser. 2021. "Methods to Analyze Negotiation Processes." *Handbook of Group Decision and Negotiation*, 39–60. Springer.

Vihma, Antto, Yacob Mulugetta, and Sylvia Karlsson-Vinkhuyzen. 2011a. "Negotiating Solidarity? The G77 through the Prism of Climate Change Negotiations." *Global Change, Peace, and Security* 23(3): 315–34.

Vogel, Jesse. 2014. "The Problem with Consensus in the UN Framework Convention on Climate Change." *Philosophy and Public Policy Quarterly* 32(2): 14–22.

Vohnsen, Nina Holm. 2017. *The Absurdity of Bureaucracy: How Implementation Works.* Manchester University Press.

Voigt, Christina, and Felipe Ferreira. 2016. "Differentiation in the Paris Agreement." *Climate Law* 6(1–2): 58–74.

Vu, Hong Tien, and Nyan Lynn. 2020. "When the News Takes Sides: Automated Framing Analysis of News Coverage of the Rohingya Crisis by the Elite Press from Three Countries." *Journalism Studies* 21(9): 1284.

Waite, Daniel. 2020. "Bloc Party: Investigating the Strategies of AILAC in the UNFCCC." Thesis, University of Reading. https://centaur.reading.ac.uk/98543/1/Waite_Thesis.pdf.

Walker, Ronald. 2004. *Multilateral Conferences: Purposeful International Negotiation.* Springer.

Walker-Crawford, David Noah. 2021a. "Climate Change in Court: Making Neighbourly Relations in a Warming World." PhD diss., University of Manchester. https://www.research.manchester.ac.uk/portal/en/theses/climate -change-in-court-making-neighbourly-relations-in-a-warming-world(0069bbad -baob-4a3b-aad8-cd56bb0bd0ab).html.

———. 2021b. "The Moral Climate of Melting Glaciers." In *The Anthroposcene of Weather and Climate: Ethnographic Contributions to the Climate Change Debate*, 146.

Wallbott, Linda, and Andrea Schapper. 2017. "Negotiating by Own Standards? The Use and Validity of Human Rights Norms in UN Climate Negotiations." *International Environmental Agreements: Politics, Law, and Economics* 17(2): 209–28.

Walsh, Sean, Huifang Tian, John Whalley, et al. 2011. "China and India's Participation in Global Climate Negotiations." *International Environmental Agreements: Politics, Law, and Economics* 11(3): 261–73.

Wang, Huan, and Wengying Chen. 2019. "Gaps between Pre-2020 Climate Policies with NDC Goals and Long-Term Mitigation Targets: Analyses on Major Regions." *Energy Procedia* 158:3664–69.

Wapner, Paul Kevin. 1996. *Environmental Activism and World Civic Politics.* SUNY Press.

Warlenius, Rikard. 2018. "Decolonizing the Atmosphere: The Climate Justice Movement on Climate Debt." *Journal of Environment and Development* 27(2): 131–55.

Warner, Koko. 2012. "Human Migration and Displacement in the Context of Adaptation to Climate Change: The Cancun Adaptation Framework and Potential for Future Action." *Environment and Planning C: Government and Policy* 30(6): 1061–77.

Warner, Koko, Kees Van der Geest, Sönke Kreft, et al. 2012. *Evidence from the Frontlines of Climate Change: Loss and Damage to Communities Despite Coping and Adaptation.* UNU-EHS.

Warren, Kay B. 2002. "Toward an Anthropology of Fragments, Instabilities, and Incomplete Transitions." In *Ethnography in Unstable Places: Everyday Lives in Contexts of Dramatic Political Change,* 379–92.

Watts, Joshua. 2020. "AILAC and ALBA." In *Coalitions in the Climate Change Negotiations,* ed. C. Klöck, Paula Castro, Florian Weiler, and Lau Øfjord Blaxekjær, 156–74. Routledge, 2020.

Watts, Joshua, and Joanna Depledge. 2018. "Latin America in the Climate Change Negotiations: Exploring the AILAC and ALBA Coalitions." *Wiley Interdisciplinary Reviews: Climate Change* 9(6): e533.

Watts, Mark. 2017. "Cities Spearhead Climate Action." *Nature Climate Change* 7(8): 537–38.

Weikmans, Romain, and J. Timmons Roberts. 2019. "The International Climate Finance Accounting Muddle: Is There Hope on the Horizon?" *Climate and Development* 11(2): 97–111.

Weikmans, Romain, J. Timmons Roberts, Jeffrey Baum, Maria Camila Bustos, and Alexis Durand. 2017. "Assessing the Credibility of How Climate Adaptation Aid Projects are Categorised." *Development in Practice* 27(4): 458–71.

White, Stephen K. 2000. *Sustaining Affirmation: The Strengths of Weak Ontology in Political Theory.* Princeton University Press.

Whitington, Jerome. 2013. "Fingerprint, Bellwether, Model Event: Climate Change as Speculative Anthropology." *Anthropological Theory* 13(4): 308–28.

———. 2016. "What Does Climate Change Demand of Anthropology?" *PoLAR* 39:7.

Williams, Marc. 2005. "The Third World and Global Environmental Negotiations: Interests, Institutions, and Ideas." *Global Environmental Politics* 5(3): 48–69.

Willox, Ashlee Cunsolo. 2012. "Climate Change as the Work of Mourning." *Ethics and the Environment* 17(2): 137–64.

Winkler, Harald. 2020. "Putting Equity into Practice in the Global Stocktake under the Paris Agreement." *Climate Policy* 20(1): 124–32.

Winkler, Harald, Brian Mantlana, and Thapelo Letete. 2017. "Transparency of Action and Support in the Paris Agreement." *Climate Policy* 17(7): 853–72.

Wisner, Ben, Piers Blaikie, Terry Cannon, et al. 2014. *At Risk: Natural Hazards, People's Vulnerability, and Disasters*. Routledge.

WNN. 2020. "Russia and Bangladesh Expand Nuclear Cooperation." *World Nuclear News*. March. https://world-nuclear-news.org/Articles/Russia-and-Bangladesh -expand-nuclear-cooperation.

Wong, Poh Poh. 2011. "Small Island Developing States." *Wiley Interdisciplinary Reviews: Climate Change* 2(1): 1–6.

Wood, James C. 1995. "Intergenerational Equity and Climate Change." *Georgetown International Environmental Law Review* 8:293.

Yamin, Farhana. 1998. "Climate Change Negotiations: An Analysis of the Kyoto Protocol." *International Journal of Environment and Pollution* 10(3–4): 428–53.

Yamin, Farhana, and Joanna Depledge. 2004. *The International Climate Change Regime: A Guide to Rules, Institutions, and Procedures*. Cambridge University Press.

Yarrow, C. H. Mike. 1978. *Quaker Experiences in International Conciliation*. Yale University Press.

Yates, Frances. 2011. *The Art of Memory*. Random House.

Younus, Md Aboul Fazal, and Md Alamgir Kabir. 2018. "Climate Change Vulnerability Assessment and Adaptation of Bangladesh: Mechanisms, Notions, and Solutions." *Sustainability* 10(11): 4286.

Zhang, Yong-Xiang, Qing-Chen Chao, Qiu-Hong Zheng, et al. 2017. "The Withdrawal of the US from the Paris Agreement and Its Impact on Global Climate Change Governance." *Advances in Climate Change Research* 8(4): 213–19.

Zhou, Chen, Mark D. Zelinka, Andrew E. Dessler, et al. 2021a. "Greater Committed Warming after Accounting for the Pattern Effect." *Nature Climate Change* 11(2): 132–36.

Index

Abdulla, Amjad, 62–64
Abeysinghe, Achala, 66–67,
 73–74
accountability, 119–21, 174–75
Action Aid, 56
Action Aid Bangladesh, 168
activism: by ADP, 17; in bureaucracy, 96–97;
 by CAN, 12, 14–15, 93–94, 186n8, 191n1; at
 COP, 93–94; by DCJ movement, 45, 84,
 186n8, 191n1, 194n1; in geopolitics, 18–24;
 by Haq, 83–89; institutions and, 19–20;
 justice in, 21; for LDC, 53–55; in marginalization, 12, 25–28; monitoring, 35; for
 NDCs, 146–51; Nigera Kori in, 40; to
 police, 36–37; protests, 26–27; racial justice, 168–69; with UNFCCC, 79–80;
 in US, 171–72; by We Are Still In Coalition, 66
adaptation, 14, 18–19, 31–32, 51, 71–72, 143,
 157, 159–60
Ad Hoc Working Group on Paris Agreement (APA): agenda items to, 140–48, 152;
 bureaucracy in, 128, 132; at COP, 104–5,
 126–28; NDCs and, 133–34; PA and, 17,
 22, 53, 68, 101, 134–35, 141
ADP. *See* Durban Platform for Enhanced
 Action
Africa, 28–29, 43, 63, 151, 160, 167–68, 188n5.
 See also specific countries
African American community, 167
African Group of Negotiators, 46–48, 54,
 63–64, 76–77, 112, 160, 164

agenda items, 127, 132, 133–34, 140–48,
 152
ALBA. *See* Bolivarian Alliance for the
 Peoples of Our America
Ali, Mirza Shawkat, 51–52, 54–55, 56, 91–92,
 176
Alliance of Small Island States (AOSIS):
 climate change to, 155, 159–60, 163–64,
 169; in geopolitics, 45–46, 86–87, 105,
 126, 147–48; SIDS-AOSIS, 51, 64, 68, 75;
 UNFCCC and, 62–64
anthropology, 8–10, 15, 18–20
AOSIS. *See* Alliance of Small Island
 States
APA. *See* Ad Hoc Working Group on Paris
 Agreement
Arab Group, 54–55
Arab League, 47
ARAMCO, 47
Asia, 28, 43, 151, 167–68. *See also specific
 countries*
Association of Latin America and the
 Caribbean (AILAC), 47
attribution, 156, 169–71
Auffhammer, Maximilian, 170
Australia, 89–90, 147
Avrill, Marilyn, 85, 89

Baashan, Sarah, 128, 133, 140–41, 143–44,
 149–50
Bakhtin, Mikhail, 9
Bandung Conference (1954), 44

Naveeda Khan is Associate Professor of Anthropology at Johns Hopkins University. She sits on the board of the JHU Center for Islamic Studies, and serves as affiliate faculty for the JHU Undergraduate Program in Environmental Science and Studies. She is the author of *Muslim Becoming: Aspiration and Skepticism in Pakistan* (Duke, 2012) and *River Life and the Upspring of Nature* (Duke, 2023) and editor of *Beyond Crisis: Re-evaluating Pakistan* (Routledge, 2010).

Printed in the USA
CPSIA information can be obtained
at www.ICGtesting.com
JSHW021323110324
58992JS00007B/251